Stone in Building

This book is dedicated to fellow-members of the Standing Joint Committee on Natural Stones, for their help, encouragement and knowledge willingly given, and to the late Donovan Purcell, first chairman of the Committee.

Stone in Building

its use and potential today

John Ashurst and Francis G. Dimes

The Architectural Press Ltd : London

Authors

JOHN ASHURST, D Arch RIBA (left) is a research architect with the Directorate of Ancient Monuments and Historic Buildings, DOE, and chairman of the Standing Joint Committee on Natural Stones.

FRANCIS G. DIMES, BSc FGS (right) is a principal scientific officer with responsibility for the national collections of building and decorative stones held by the Geological Museum which is part of the Institute of Geological Sciences.

With contributions from:
PHILIP BURTON
ARTHUR FORDHAM
J. V. BORTHWICK
D. MAXTEAD JONES
AUSTIN SILCOX CROWE
TED BEDFORD
REG WOOD
M. J. BOWLEY
C. A. PRICE
IAN CLAYTON

First published in book form in 1977 by The Architectural Press Ltd: London
ISBN: 0 85139 607 0

All rights reserved. No part of this publication may be reproduced, stored in a retrieval system, or transmitted, in any form or by any means, electronic, mechanical, photocopying, recording or otherwise, without the prior permission of the publishers. Such permission, if granted, is subject to a fee depending on the nature of the use.

© John Ashurst and Francis G. Dimes 1977
Printed litho in Great Britain by W & J Mackay Limited, Chatham

Contents

Scope and form of the handbook vi

1 **Geology**
- TS1 Introduction and classification 1
- TS2 Igneous rocks 6
- TS3 Sedimentary rocks 10
- TS4 Metamorphic rocks 18

2 **Quarrying**
- TS1 Nature and location of quarries 22
- TS2 Quarrying methods 24

3 **Processing**
- TS1 Workshop practices 28
- TS2 Site works and cladding 33

4 **Training**
- TS1 Training: organisation and apprenticeship 36

5 **Maintenance**
- TS1 Cleaning and surface treatments 40
- TS2 Diagnosis and repair methods 51
- TS3 Recent developments in in-depth treatments 58

6 **Specification**
- TS1 Testing porous building stone 62
- TS2 Notes and specification 65
- IS1 Thermal transmittance 68

7 **Stone in use**
- DS1 Bonded ashlar walls 71
- DS2 Solid stone walls and windows 72
- DS3 Rubble walling 73
- DS4 Masonry walls ashlared with stone 74
- DS5 Masonry walls ashlared with stone 75
- DS6 Stone faced concrete walls 76
- DS7 Fixings for ashlar facings over 75 mm thick 77
- DS8 Joggle joints and fixings for ashlar facings under 75 mm thick 78
- DS9 Copings and soffits 79
- DS10 Slate roofing 80
- DS11 Stonework drawing 81
- DS12 Stonework drawing 82
- DS13 A modern stone building 83
- DS14 A traditional stone building 84
- DS15 Buildings in stone 85
- DS16 Buildings in stone 86
- DS17 Decorative use of stone 87
- DS18 Decorative use of stone 88

8 **Hard landscape in stone** 89

9 **Stone substitutes** 94

Appendix 100

Acknowledgements 101

Index to stones 102

General index 104

Scope and form of handbook

Introduction
It is a long time since most architects thought in stone and only comparatively recently have a few with an understanding and appreciation of the material taken to using it imaginatively. Stone has many applications besides the traditional ones and as simple cladding slabs. With a magnificent architectural heritage in stone there is a danger of our consigning the material to the past, of thinking of stone mainly in terms of restoration. Little attention is paid to design in stone in architectural schools and most architects are not now capable of designing in it correctly and economically.

A major reason for this neglect has been an ill-informed idea of the high cost of stone. It can be expensive, but how many architects have compared it with other materials for a specific job, discussed its use with a supplier at an early design stage, or considered its low maintenance cost?

There are ways in which stone can be used without enormous cost, and its availability, especially when compared with petroleum-derived materials, makes it increasingly attractive. Stones are often readily available and need little energy for extraction and processing.

Nor is the impression that the stone industry is in decline any longer correct. Changes in organisation, training and promotion are all beginning to have effect.

Improved mechanisation and the use of waste products will contribute increasingly to bringing down the relative price of stone. The industry too is well provided with technical information: worldwide research is going on into maintenance, preservation and testing, and more is known about the weathering and durability of stone than about those of almost any other material.

Handbook format
The sections have been prepared by a number of experts with the aid of several others, not directly contributing, whose invaluable advice is gratefully acknowledged. The handbook is intended to give comprehensive coverage of stone as a building material and emphasises particularly previously neglected areas.

The first four sections: Geology; Quarrying; Processing and site works; and Training, provide the background information on the nature of stone and the forms of stonework that can be provided. Section five describes the current practice of Repair and maintenance. Section six, Specification, contains information on stone testing and specifying stone, and data on thermal insulation properties.

After these sections, which consist of technical studies, section seven provides traditional and recent details for stone. This is followed by two further sections dealing, respectively, with hard landscape in stone and with stone substitutes. The handbook is completed by the Index to Stones and the General Index.

Detail sheets
The technical studies and information sheets follow usual AJ style. Detail sheets are unique to this handbook and provide working details of traditional and recent construction using limestone, sandstone, granite, marble and slate.

Geology Technical Study 1
Introduction and classification

This first technical study by FRANCIS G. DIMES introduces the stones used for building, how they are formed and the systems of joints to be found in rocks in the ground.

The criteria for stone for building
1.01 Dimension stone, an increasingly used term for a rock that can be cut and worked to a specific size or shape for use in building must be obtainable in blocks large enough for the purposes in mind, be free from fractures, tough and devoid of minerals which may break down chemically or by weathering. Hardness is not necessarily a requisite although where a stone is used for steps or flooring, resistance to abrasion is an essential quality.
No two blocks of stone, even if quarried side by side, are absolutely identical. The differences may not be discernible and of no practical importance; but they may be substantial. Each block must be considered individually. The differences may contribute to the beauty of the stone which must be regarded as a material of variety requiring an individual treatment so that it is used to the best advantage.

Geological history
1.02 The geological history of rocks used by man ranges over 3800 million years. An originally molten earth cooled and a solid crust was formed. Much of this was weathered away by rain, wind, heat and cold, the resulting fragments being carried away by transporting agents to be redeposited, mostly in the seas. Much has been remelted and later again solidified. Much has been subjected to intense pressures and stresses, to heat and to chemical changes. As a result of these complex forces a vast variety of rocks has been produced differing markedly in appearance and in nature.

Groups of rocks
1.03 Great Britain probably has a greater variety of rock in a smaller area than has any comparable part of the world. They can be placed in three main groups.
Igneous rocks—once molten these were formed by the consolidation of *magma*, the hot, more or less fluid rock material originating deep within or beneath the earth's crust.
Sedimentary rocks—ancient sediments originally deposited on the beds of seas, lakes and rivers, or on older land surfaces and now to a greater or lesser extent compacted and naturally cemented.
Metamorphic rocks—distinctive new rock types produced by the recrystallisation of pre-existing rocks (with or without a change of composition) by heat, pressure and chemical fluids acting separately or in combination, while the rock remained essentially solid, **1**.

Composition
1.04 All rocks are aggregates of minerals (ie natural inorganic substances with symmetrical crystal forms reflecting internal atomic structures, which have definite chemical composition). Only about 25 minerals, either singly or in association, make up the physical bulk of rocks used for building.[1]

Formation and classification of rocks

Igneous
2.01 The rocks within the earth normally are solid. They melt, forming magma, when there is a release of pressure from the overlying rocks or an addition of other materials.
2.02 In geologically favourable conditions magma may rise through the crust, and may be poured out as lava during a volcanic eruption, cool rapidly and form glassy, non-crystalline or finely crystalline rocks. The resultant rock is called a volcanic or *extrusive* igneous rock. If cooled within the crust it forms *plutonic* or *intrusive* igneous rock. Major intrusions within the crust (which may be of many cubic kilometres blanketed by overlying rock formations) cool slowly to form coarsely crystalline rocks in which the different minerals may be individually recognised, **2**.
2.03 Igneous rocks are essentially assemblages of silicates and the proportion of oxide of silicon (silica, SiO_2) present may be used as a basis of classification. The most siliceous igneous rocks are known as *acid* (a purely chemical expression) and contain quartz, the crystal form of silica. Rocks which are poor in silica are known as *basic* and *ultra-basic*. A classification based both on the percentage of silica present, and the position of emplacement, is shown in table I.

Sedimentary
2.04 The raw materials of sedimentary rocks are sediments produced by the geological processes of weathering, erosion and sedimentation, which originated, often by many derivations, from other rocks. For example constituents of granite, an acid plutonic igneous rock containing quartz, feldspar and mica, weather in different ways.
Quartz, hard and chemically resistant, becomes broken into smaller fragments. Mica, extremely resistant to weathering, is broken down into flakes of sub-microscopic size. Feldspar, especially when acted upon by slightly acid water, breaks

Table I Classification of igneous rocks based on percentage of silica and position of emplacement					
	Silica, per cent	65-75 per cent *Acid*	55-65 per cent *Intermediate*	45-55 per cent *Basic*	35-45 per cent *Ultrabasic*
Position of emplacement	Volcanic (glassy or fine grained)	Pumice Obsidian Rhyolite	Andesite	Basalt	
	Minor intrusion (fine or medium grained)	Quartz-porphyry	PORPHYRY	Dolerite	
	Plutonic (medium or coarse grained)	GRANODIORITE GRANITE	DIORITE SYENITE	GABBRO	Some SERPENTINITE

Note: Those igneous rocks which have been used on any scale for building or decorative purposes are shown in capitals.

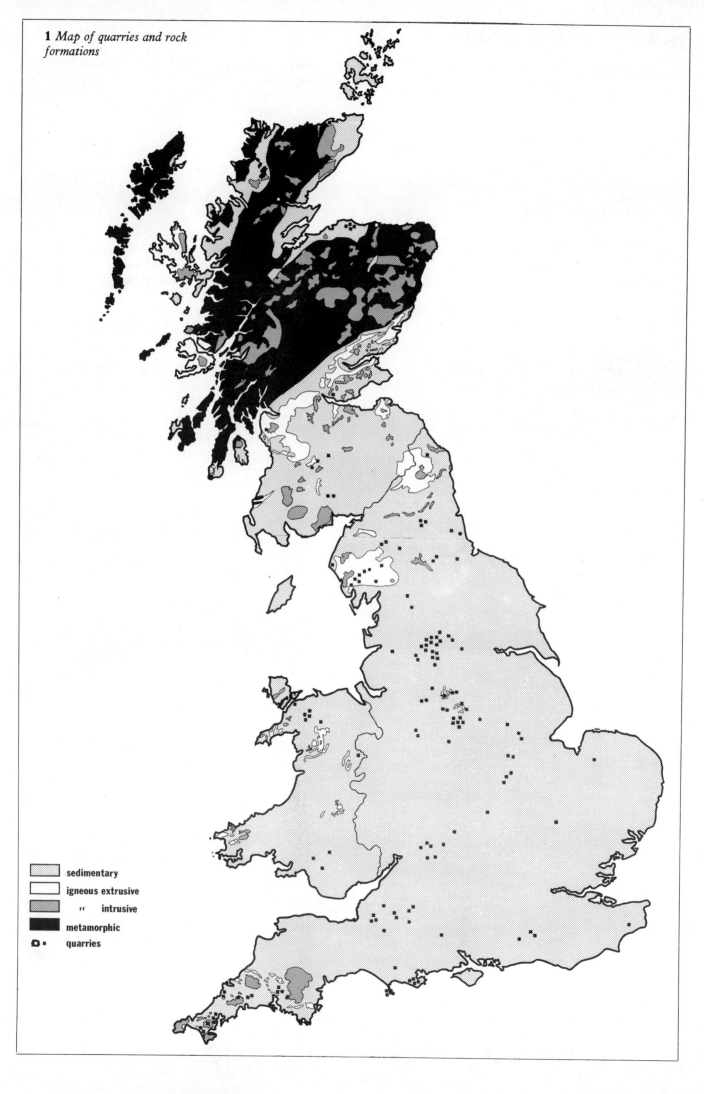

1 *Map of quarries and rock formations*

Table II Classification of sedimentary rocks

Rudaceous rocks (rubbly rocks mainly composed of large fragments of older rocks)
Breccia	A rock consisting of broken, angular, unworn fragments held together by natural cement
Conglomerate	A rock consisting of rounded fragments held together by natural cement

Arenaceous rocks (sandy rocks)
SANDSTONE	A bedded rock composed of grains of quartz sand naturally cemented together
GRITSTONE	A sandstone with angular and usually coarse grains (a coarse greywacke)
Greywacke	A fine grained compact sandy rock, with badly sorted angular fragments and a finer grained matrix of clay
ARKOSE	A feldspathic sandstone or grit, medium to coarse grained and containing over 25 per cent feldspar
Quartzite	A rock composed almost entirely of closely fitting quartz grains, naturally cemented by secondary quartz

Argillaceous rocks (clayey rocks)
Clay	A fine grained sediment of minute flaky minerals
Shale	A laminated compacted argillaceous rock

Calcareous rocks (carbonate rocks, mainly composed of calcium and magnesium carbonates)
LIMESTONE	A bedded rock consisting essentially of calcium carbonate
OOLITE	A limestone characterised by abundant small spheroidal calcareous grains
Pisolite	Resembling oolite but containing spheroidal bodies about the size of a pea
Tufa	A calcareous deposit from saturated limey waters
TRAVERTINE	Large masses of calcareous sinter, similar to tufa but more compact
MAGNESIAN LIMESTONE	A limestone containing magnesium carbonate
Dolomite	Composed mainly of the mineral dolomite, the double carbonate of calcium and magnesium

Table III Classification of metamorphic rocks

Sedimentary origin
SLATE	Pelitic rocks: from argillaceous sediments
Mica-schist	Psammitic rocks: from arenaceous sediments
Quartzite	
	Semi-pelitic rocks: from impure arenaceous sediments
	Metamorphosed greywacke
MARBLE	Calcareous rocks: from calcareous sediments

Igneous origin
Schist, gneiss	Basic rocks: from dolerites and basalts
Schist, gneiss	Acid rocks: from granites and allied rocks

Note: Those sedimentary and metamorphic rocks which have been used on any scale for building or decorative purposes are shown in capitals.

down to give soluble salts and releasing the quartz and mica grains.

2.05 Basic igneous rocks weather in much the same manner; more soluble material and sometimes more clay are produced but no quartz grains.

2.06 Although these weathering products may remain stationary to blanket the bedrock from which they were formed, more often they are transported. Running water and the seas not only sort these materials one from the other but also concentrate similar materials together. Small insoluble flakes of mica are deposited as *muds* in layers ('beds'), the quartz grains as beds of *sand* and the salts are added to the sea where some may be chemically precipitated to form, for instance, *lime mud*. Sooner or later the sediments are converted into sedimentary rocks by dewatering, compaction, welding of their constituent particles and by natural cementation of the constituent grains, 3, 4, 5.

2.07 Sedimentary rocks may be classified according to grain sizes and the nature of the grains, as in table II.

Metamorphic

2.08 Rocks within the earth's crust may be subjected to heat, pressure and chemical change resulting from crustal deformations. The original mineral matter may recrystallise or new stable minerals may be produced.

2.09 There are two categories of metamorphism. First, that caused by heat from an igneous intrusion, known as *thermal* or *contact metamorphism*. Second, that characteristically associated with mountain building and accompanied by deformation, known as *regional metamorphism*. Any rock may be metamorphosed and produce a distinctive new rock type but most metamorphic rocks arise from the common

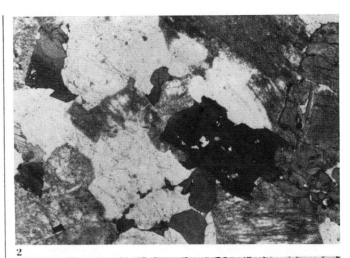

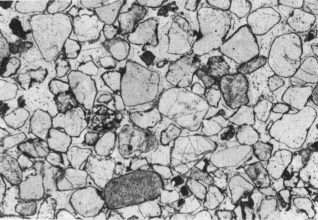

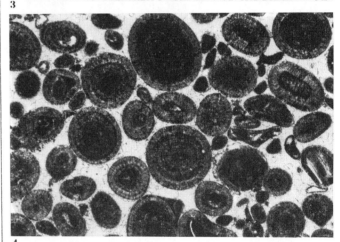

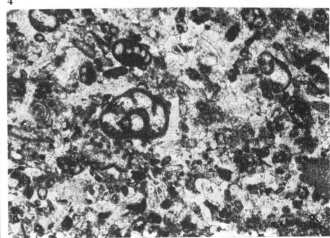

2 *Thin section of granite (field of view 3 mm).*
3 *Thin section of sandstone (field of view 1 mm).*
4 *Section of oolitic limestone (field of view 4 mm).*
5 *Section of limestone with fossil fragments (field of view 4·5 mm).*

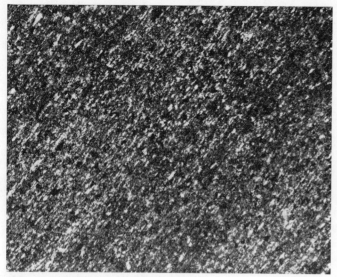

6 Section of slate (field of view 4 mm).

igneous and sedimentary rocks. Only two are widely used for building purposes: marble and slate.

2.10 *Thermal metamorphism* is a matter of simple baking. When heated and confined by overlying rocks, the particles of calcium carbonate ($CaCO_3$) and the fossils in a limestone are gradually recrystallised into roughly similar sized crystals of calcite. The original limestone is converted into a marble. In true marble all traces of fossils are destroyed. Limestone with little other mineral matter is converted into a pure white granular marble. Other minerals, if present, change by chemical reactions into new minerals which give the colour and 'figuring' of many marbles.

2.11 *Regional metamorphism* is widespread and takes place in regions of mountain building where rocks are subjected to continuous stress. The flaky minerals present in clay tend to re-orientate at right angles to the pressure. Recrystallisation also takes place and small new flakes of minerals (principally mica and chlorite) grow at right angles to the pressure. The new rock (now called a slate) has had impressed on it a parallel orientation of planar minerals. It has a grain known as *slaty cleavage* along which the rock may easily be split, **6**. The direction of this cleavage commonly has no relation to the original bedding of the sediment. Metamorphic rocks exist in large numbers and any rock may be metamorphosed in one of two ways. The classification in table III is based on the original nature of the rock.

3 Joints
Joint system

3.01 A rock must yield blocks large enough for building purposes; the size is controlled by joints. These are fractures or parting planes which separate a mass of rock (not to be confused with a fault which is a fracture within a rock mass along which the rocks on one side have moved relatively to those on the other). Groups of parallel joints form a joint set and intersecting sets of joints form a joint system. Joints and joint systems originate in different ways: some form during hardening or cooling of rock, others are caused by tensional forces in the earth's crust.

Igneous rocks

3.02 During a volcanic eruption, magma migrates to the surface and pours out as lava, **7**. When this cools, it shrinks and may crack in a honeycomb pattern. These cracks also grow at right angles to the cooling surface and the lava sheet is divided into hexagonal columns. Flat lying lava sheets therefore have a vertical joint system. Cross joints are developed to a lesser extent and the columns are divided into shorter lengths.

3.03 Minor intrusions, such as sills (roughly horizontal) and dykes (steeply inclined) show similar but less well developed columnar jointing, at right angles to cooling surfaces. Major plutonic intrusions show three sets of joint systems, caused by contraction on cooling and stresses caused by overlying rocks. Two sets are vertical and roughly at right angles while the third, sheet jointing, is horizontal, commonly following the contours of the surface. If exposed at the surface, weathering opens up these joints. Joints in granite can be widely spaced. Cleopatra's Needle, London, of granite from Egypt is recorded as one of the longest unjointed stones in the world.

Sedimentary rocks

3.04 The product of a single episode of sedimentation is a *bed*. It is bounded top and bottom by a bedding-plane. Beds may be regular with parallel bedding planes, or irregular with bedding planes of successive beds meeting at an angle, **8**.

3.05 A bedding plane is a primary structure of sedimentary rocks. It indicates a pause in sedimentation during which the surface of the bed may be rain-pitted or ripple marked,

7 Lava flows, Iceland. Lava is the name for both molten and solid form whatever its age.
8 Cliff face of sedimentary rock with close bedding and well-marked bedding planes.

may dry and show sun cracks, or may lie quietly under water with very fine material settling on it to form a thin layer. A series of beds built up is called bedding or stratification.

3.06 As a sediment dries out, shrinkage joints develop, limited by the bedding planes. Sedimentary rocks frequently show two joint sets approximately at right angles. Strongly developed and persistent joints which frequently are found to cross several beds, are termed master joints and these may determine the manner in which sedimentary rocks are quarried. Well jointed rocks are more easily extracted, **9**.

Metamorphic rocks

3.07 Metamorphic rocks are the products of change by heat, stress and chemically active fluids. Low temperature metamorphism produces fine grained rocks while high pressure and temperature over a long period of time produce coarse textured rocks. The characteristic texture of metamorphic rocks is foliation, the parallel arrangement of minerals.

3.08 Coarse grained metamorphic rocks in which the mineral grains are orientated are said to show a schistose texture. The foliation of a schist, the most common metamorphic rock, is seen as discontinuous layers. If subjected to strong stress marble will develop rough schistosity. Segregation of the minerals into rough bands is termed a gneisose texture. A gneiss, the product of high grade regional metamorphism, is coarse grained, with different minerals in bands, and has poor foliation.

3.09 In general, because of the ease with which they will part along the foliation, metamorphic rocks have not been used on any scale for building purpose, slate and marble being the two outstanding exceptions.

4 The age of rocks

4.01 If a house is described as Georgian it may immediately be concluded that it was built after houses called Elizabethan and before buildings of the Victorian period. The description implies a relative age not expressed in years and specific characteristics. The same general principles may be applied to rock formations. Sedimentary rocks were deposited layer upon layer usually with the younger layer lying on the older. A sequence of geological events can be determined and placed on a time scale.

4.02 The rate of sediment deposition was not constant in area or in time. Subsequent erosion may also have removed much of the thickness of the rocks. In many places the rocks have been folded, faulted and even inverted. But by studying the relationships a geological column has been constructed, though at no one place on earth is the complete succession present. This column, table IV, is the fundamental reference document of the geologist. A rock described as Jurassic was laid down after Triassic rocks and before those of the Cretaceous period. It has been given a relative age. The description also implies certain forms of fossils and other characteristics.

4.03 In recent years it has been realised that the decay of unstable elements may be used to give rocks an absolute date in terms of years. Radioactive elements decay at a constant rate and isotopes of three of them—uranium, potassium and rubidium—are used as geological clocks. These absolute ages have been added to the geological column in table IV.

Reference

1 A list of commonly found minerals is given in *Glossary of terms for stone used in building*, British Standard 2847, pp8-9, BSI, 1957.

9

Table IV The geological column		
Era	Period	Age in millions of years
Caenozoic	Quaternary	1
	Tertiary	65
Mesozoic	Cretaceous	136
	Jurassic	190
	Triassic	225
Palaeozoic	Permian	280
	Carboniferous	345
	Devonian	395
	Silurian	430
	Ordovician	c500
	Cambrian	c570
	Precambrian	c4500

9 *A Portland stone quarry. Well bedded stone with marked joint system.*

Geology Technical Study 2

Igneous rocks

This study by FRANCIS G. DIMES describes geologically the available igneous building stones, giving examples of appropriate use.

The naming of rocks

In the stone trade many stones are given names which do not accord with geological usage.

For example, almost any stone may be called a granite; 'Petit Granit' a dark limestone from Belgium; 'Ingleton Granite', a grit from Yorkshire; 'Blue Pearl Granite', a syenite from Norway, are examples. The scientific name of a stone is of importance. Many years ago it was realised that names more in accordance with scientific practice would be an advantage. The properties of a rock are, to a considerable extent, functions of its mode of formation, composition and texture. If the scientific name of a rock is used, then the purchaser should get the type of rock best suited for his purposes.

1 Granite rocks

1.01 Granite is the only igneous rock which has been widely used for building and ornamental work, providing a great variety of stones ranging in colour from pale grey to blue-grey and red.

1.02 Geologically, a granite is a strictly defined rock. Used here the name is extended to include medium- or coarse-grained igneous rocks which contain free quartz and whose other main constituent minerals are feldspar and mica.

Granites of Devon and Cornwall

1.03 Granites from Devon and Cornwall have been in use since the Bronze Age. Although on a geological map the granite masses of Dartmoor, Bodmin, St Austell, Carnmenellis, Land's End and the Isles of Scilly are shown as separate outcrops, they join at depth to form one elongated mass. Within the mass are several varieties of granite; the most widespread contains large well-formed ('porphyritic') feldspar crystals. A more uniform and finer-grained variety is found on the edges of the Land's End and the Bodmin masses.

1.04 All the granites of the area contain quartz, feldspar* (orthoclase) which is often large and porphyritic (but not always so), and predominantly black (biotite) and white (muscovite) mica. Other minerals may be present in far lesser amount. Tourmaline is a common 'accessory' mineral.

1.05 All granites in production in south west England† are silver-grey in colour, though at one time a pink stone was produced. Granites from south-west England have been used widely, **1**, for examples the TUC building, the Thames Embankment, the foundations of the Houses of Parliament and the Eddystone lighthouse.

1.06 In the St Austell mass, feldspar within the granite has been converted into kaolin (china clay). At Luxulyan the granite has been tourmalinised and an attractive pink and black stone named luxullianite is found. It was used for the tomb of the first Duke of Wellington in St Paul's Cathedral.

Shap granite

1.07 Widely known and frequently employed decoratively, especially in late Victorian times, Shap granite has an outcrop of only 7·5 km². In colour it ranges from grey to warm brownish-red. The stone is recognised by its characteristic large prophyritic crystals of pink coloured feldspar up to 50·8 mm long, set in a finer ground-mass of smaller feldspar crystals, black mica and quartz, **2**. Two distinct shades known as 'light Shap' and 'dark Shap' are supplied, which grade into one another. The 'dark' variety is found in bands some 1 to 4 m wide on each side of the master joints. Commonly, the stone shows dark coloured patches known by quarrymen as 'heathen' or, less frequently, as 'foreigners', and by the geologist as xenoliths. They are fragments of earlier formed rocks caught up in the magma before it solidified. 'Heathen' add additional character to the stone and are not flaws which in any way affect the quality.

---

* See glossary on page 9

† A list of quarries producing dimension stone is given in *Natural stone directory*, Ealing Publications Ltd, 73a High Street, Maidenhead, Berkshire, 1977

1 *New Scotland Yard (architects Chapman Taylor & Partners), an example of Cornish granite.*
2 *Monolithic bollards of Shap granite set round St Paul's.*

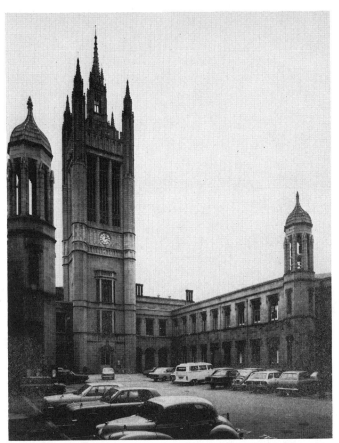

3 Marischal College, Aberdeen University, 'a poem in stone', the second largest granite building in the world (after El Escorial, Spain), is of Kemnay granite.

Scottish granite

1.08 Large masses of granite occur in the Southern Uplands, the Grampian Highlands, the Northern Highlands and in the Western Isles. The granites range in age from Precambrian to Tertiary. Of prime importance are those intruded during the Caledonian mountain building episode which took place 400 million years ago late in the Silurian and early Devonian periods. These granites are typified by Aberdeen, the 'granite city', which stands on the eastern edge of a series of related granite masses stretching westwards through Tarland to Ballater and to Loch Etive. The masses of Peterhead and Strichen to the north of Aberdeen belong also to this series.[1]

Peterhead granite

1.09 Peterhead granite is usually coarse-grained and dark flesh-coloured, consisting mostly of quartz and feldspar (orthoclase). It has been widely used not only for engineering works but also for decoration. The old prison quarry at Peterhead was recently (1973) re-opened for dimension (sized) stone and produced blocks up to 1 m^3 but by 1976 had ceased production. A fine-grained, dark grey variety, with irregular-shaped porphyritic white feldspar crystals, was quarried at Cairngall and known as Blue Peterhead. It was obtainable in large blocks and extensively used for decorative work, the best-known being the 30 tonne block for the sarcophagus of the Prince Consort. St George's Hall, Liverpool, is renowned for eight pillars, each a monolith 6 m tall, of this granite.

Aberdeen granite

1.10 Aberdeen granite comes from many quarries in the area, some of them world famous. Rubislaw quarry, said to date back to 1741, produced a medium-grained, dark bluish-grey stone containing xenoliths. In it, abundant biotite mica flakes have a definite orientation, so the face selected for polishing should be cut across the edges of the mica. If the cut is parallel with the mica flakes, the polished face will soon loose its appearance because of the flaking or 'cleavage' of the mica. This consideration applies to all granites with orientated micas. Rubislaw granite was used for Waterloo Bridge (in part, 1817) in Portsmouth Docks and once was exported to many countries.

Kemnay granite

1.11 Kemnay granite is medium-grained, light-grey and contains orientated biotite mica. Pink granite, identical except for colour with the grey, occurs as bands on either side of the major joints, **3**.

Correnie granite

1.12 This occurs in two types. The most abundant is a distinctive light or dark pink granite, fairly coarse-grained with a streaked structure, wide-jointed and available in large blocks. The other, a grey quartz-diorite which occurs as patches or bands, is often so firmly welded to the pink granite that a block from the quarry may be half pink and half grey. Pink Correnie granite was used for the Municipal Buildings, Glasgow.

Ross of Mull granite

1.13 This was one of the earliest to be quarried in western Scotland. It varies in colour from pale to deep red, contains predominantly black biotite mica and is coarse and fairly even grained. The very wide spacing of the joints allowed large blocks to be extracted. Large blocks were used as pedestals to the figures on Holborn Viaduct and for the piers of Blackfriars Bridge, London.

1.14 A microgranite of unusual composition comes from Ailsa Craig in the Firth of Clyde. It contains the mineral riebeckite. The stone is known in many parts of the world for it has been quarried since 1864 for curling stones. It was also used for high quality setts.

Channel Isles granite

1.15 The Channel Isles, particularly Jersey and Guernsey, have for long supplied Great Britain with granite, much of which was used for setts. At present the Ronez quarries, St John, Jersey, yield a light whitish-pink coloured stone with scattered dark mica flakes.

Imported granite

1.16 Aberdeen was for many years one of the main centres for granite-processing and many imported stones worked in the city have been given British names. 'Balmoral granite', for example, a coarse- to medium-grained, red granite with black patches, is quarried in Finland. 'Bon Accord Red', medium- to coarse-grained, red in colour, with large feldspar masses is from Uthammar, Sweden. Also from Sweden are the red 'Virgo granite' and 'Rose Swede', with dark red feldspar crystals and with distinctive blue to purple coloured quartz. Norway has produced 'Grey Royal granite', fine- to medium-grained and grey in colour, which was used for the Stock Exchange, Manchester and in the Royal Liver Building, Liverpool.

1.17 In recent years, Portugal has developed its granite industry. There is little variation in colour, most being light brown. Some of these stones are not granite in the strict geological sense.

1.18 Granite is found in many other countries. A number are quarried in Italy, of which Baveno granite from the Piedmont region, a delicate shade of pink, is widely used in Italy, for example for the huge columns in the main porch of Milan Cathedral. Another from Millbank, South Dakota, US, sometimes called 'Imperial Mahogany' is a dull red coloured stone, with blue coloured quartz. Stone faced buildings are becoming more common as the unsatisfactory weathering appearance of many designs in concrete becomes more evident, 4.

2 Other igneous rocks

2.01 In the UK, apart from granite, other igneous rocks have not been widely used; but locally they have been used with great effect. Syenite (see table I, TS1) is rare compared with granite and may be thought of as a granite without quartz. It occurs near Loch Assynt, Sutherland. Diorite, another intermediate igneous rock, composed of feldspar with hornblende, is of limited occurrence in the UK. Gabbro, a coarsely crystalline plutonic basic igneous rock, is well known as a rock type as it forms the Cuillins of Skye, but it has been used only locally for building.

2.02 Serpentinite, an ultrabasic rock composed mainly of the mineral serpentine with a supposed resemblance to the skin of a serpent, was once wrought extensively in the Lizard, Cornwall. It ranges in colour from dark red often mottled black, green with lighter green mottling to very light yellow. It is a soft stone unsuitable for outside work, soon assuming a dingy appearance. It was widely used as an interior ornamental stone, particularly in ecclesiastical work.

2.03 Polyphant stone, an allied rock, is grey-green in colour with white specks and yellowish-brown blotches. It is soft and easily carved. A fine example of its use is Archbishop Temple's tomb in Canterbury Cathedral.

2.04 Porphyry, the name derived from the Greek word for purple, was quarried in Egypt by the Romans as *Porphyrites lapis*. It consists of large minerals with well developed crystal forms in a markedly finer ground mass. The large crystals ('phenocrysts') of a porphyry include either quartz (a quartz-porphyry) or feldspar, or both. Many porphyries, locally called elvan, occur in Devon and Cornwall associated with the granite masses. Among the elvans, pride of place is taken by Pentewan stone, pale buff grey in colour, fine-grained with porphyritic feldspar and widely used for many churches in south-west England.

2.05 Dolerite, an intrusive rock type common in north-east England and the Midland Valley of Scotland has rarely been used in buildings, but where it has, it provides striking examples of the use of local material. Hadrian's Wall is in considerable part built of and on the Whin Sill, a quartz-dolerite intrusion.

2.06 Basalt, probably the commonest igneous rock in the world, makes up the Giant's Causeway, Co Antrim, the Isle of Staffa, Campsie Fells and Kilpatrick Hills, Glasgow and many other areas. The Giant's Causeway is noted for the hexagonal pillars which were produced by shrinkage as the originally molten lava flows solidified, 5.

Imported igneous rocks

2.07 Granodiorite, although commonly called granite, has more of the variety of feldspar known as plagioclase than orthoclase and therefore is not true granite. 'Silver White granite', from Norway, is a widely used example. It is coarse-grained and silvery-grey with black flecks in colour, used recently, for example, for Barclays Bank in Fountain Street, Manchester.

2.08 Syenite, a relatively rare rock, is widely used for shop fronts and for many public houses. Known geologically as Larvikite, from the town in southern Norway, varieties are sold with names such as 'Blue Pearl', 'Emerald Pearl', 'Light Pearl' and 'Imperial Pearl'. The mineral feldspar makes up the bulk of the rock and the crystals are commonly found intergrown. When cut and polished these crystals show a play of colours (schiller effect) peculiarly called 'butterfly wings' in the stone trade. 'Swedish Green granite' also is a syenite. It is coarse-grained and dark olive-green in colour. A green-coloured mineral, epidote (calcium aluminium iron silicate) gives it the greenish cast.

4

5

5a
4 *Rc framed office with granite faced sills and columns, (Fitzroy Robinson & Partners).*
5 *Giant's Causeway Northern Ireland: marked columnar jointing in basalt lava flow.*
5a *Public lavatories at The Causeway reflecting the geological character of the basalt columns.*

2.09 'Diamond Black granite' from South Africa, a syenogabbro in geological terms, has recently been imported. A black coloured stone with shiny laths up to 1 cm long of feldspar, it may be seen as the facing to the VIth Inn, Crown Square, Manchester. 'Ebony Black granite' from Sweden, medium- to fine-grained, very nearly even black in colour with shiny metallic flecks, is a diorite which has been widely used for shop fronts and for the podium of the Tower Hill development, London, 7.

'Bon Accord Black granite', also from Sweden and again principally used for shop fronts, is intermediate between a diorite and a gabbro. It is medium-grained, black in colour with light bluish-grey flecks. 'Black granite' from Sweden takes a high polish and is often mistaken for marble. It is a very dark, nearly black, medium-grained gabbro, used in the interior of the Ritz Hotel, London.

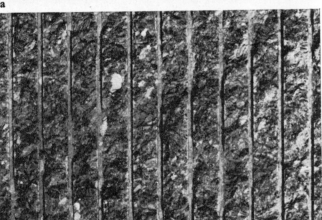

6a *Natwest Bank in King Street, Manchester in rough finish with marked vertical saw cuts.* **b** *Close-up.*
7 *Podium, Tower Hill; contrasting surface finishes.*

Glossary

For **Devonian, Precambrian, Silurian** and **Tertiary** see table IV, TS1.
Biotite: hydrous alumino-silicate of potassium, magnesium and iron (black mica).
Feldspar: a group of minerals; rock-forming alumino-silicate of potassium sodium or calcium.
Granodiorite: intermediate between granite and quartz-diorite.
Hornblende: a complex isomorphous series of silicates of calcium, iron, magnesium and other metals.
Muscovite: potash mica (white mica).
Orthoclase: potassium aluminium silicate (feldspar group).
Petrological: pertaining to stone.
Porphyritic: material containing large crystals (phenocrysts) set in a fine groundmass.
Quartz: the commonest crystalline form of silica.
Riebeckite: hydrous metasilicate of sodium and iron occurring as dark blue monoclinic prismatic crystals.
Syenogabbro: coarse grained rock intermediate between syenite and diorite in feldspar content.
Tourmaline: a complex borosilicate of aluminium together with other elements.

Reference

1 A more detailed coverage is given in J. G. C. Anderson, *Memoirs of the geological survey, Special reports on the mineral resources of Great Britain, The granites of Scotland,* HMSO, 1939.

Geology Technical study 3

Sedimentary rocks

This study by FRANCIS G. DIMES describes the two main groups of sedimentary rocks used for building: sandstones and limestones. He explains their geological form and gives examples of their use.

1 Introduction

1.01 As with igneous rocks, there is a great variety of sedimentary rocks. They were formed by the accumulation of rock material (sediments) by rivers, seas or the wind and then lithified (the process of conversion of sediments into sedimentary rocks). Because they were deposited by so many different natural agents their chemical and physical properties differ considerably. Mixtures of rock and mineral constituents are very common. In few instances does one quarry produce stone identical with that from another. Indeed, one quarry may yield a number of distinct building stones.

1.02 All, however, have a common characteristic, a layered structure ('bedding' or stratification). They form about 65 per cent of the earth's surface. Yet, despite their variety and widespread occurrence, only two big groups of the sedimentary rocks provide the major building stones. They are the arenaceous (sandstone) rocks and the calcareous (limestone) rocks. (See table II, p. 3.)

2 Sandstone

2.01 A sandstone is a *clastic* (formed from rock fragments) rock with a grain size between about 0·06 mm and 2 mm. It is usually composed predominantly of quartz grains, but rarely may be of other minerals. The grains are to a varied extent secondarily cemented by other mineral matter precipitated from ground waters circulating between the grains. The cement commonly may be siliceous, calcareous or ferruginous (iron-y). If the quartz grains are cemented with secondary silica to the extent that the rock is virtually homogenous, the resultant rock is termed a quartzite.

2.02 It is estimated that silica (the oxide of silicon) comprises 60 per cent of the earth's crust. In its mineral form, quartz, it makes up much of the sand on the beaches and it is not surprising, therefore, that sandstone is a common rock.

Mineral content

2.03 Sandstones may be pure white or coloured, the colour depending on other mineral grains present which impart special characteristics to the stone. Glauconite (a complex silicate of potassium and iron), green in colour, occurs in many sandstones and is responsible for the green tint of many of them. The name *greensand* is commonly used.

2.04 Mica is found as a constituent of some sands. If when deposited the quartz grains and mica flakes were roughly segregated with the mica flakes lying flat on the bedding planes, the rock has a fissile tendency. If such a micaceous sandstone can be split easily along the bedding planes the stone is termed a *flagstone*.

2.05 Feldspar may amount to one-third or more of the bulk of a sandstone. Rapid erosion of granites with subsequent rapid deposition does not allow the feldspar to decompose and disappear; it is sealed off from weathering. If a sandstone contains more than 25 per cent feldspar it is termed an *arkose*. Such rocks are normally of a coarse granular texture.

2.06 Dolomite (the double carbonate of calcium and magnesium), more commonly a constituent of limestones, occurs in some sandstones either as the cementing material or as discrete grains. The term dolomitic sandstone is then used.

2.07 Iron in its many compounds is often the cementing medium. It may occur also as a thin coating around each sand grain. Cementation by some iron compounds is poor and the rock is of low strength and weathers badly.

2.08 As sandstones may be composed of more than one type of constituent mineral it is not always easy to define the line of separation between them and other rock types; for instance between a calcareous sandstone and a siliceous limestone.

3 Types of sandstone

3.01 All the periods in the geological column yield sandstones. The text should be read in conjunction with table I which classifies sandstones, gives their common names, sources, characteristics and examples of their use and shows their position in the geological column.[1] Sandstones of Precambrian, Cambrian, Ordovician and Silurian age have not been used generally outside the areas in which they occur.

Devonian—Old Red Sandstone

3.02 The oldest period to have yielded stone on a large scale, the main rocks used are a series known as Old Red Sandstone which occupies a large area of Herefordshire, Monmouthshire, Brecon, Shropshire and Worcestershire. Building stones widely used around Hereford are given the general name 'Hereford Stone'.

Carboniferous

3.03 One of the most important sources of sandstones in the UK, the system is split geologically into three major divisions: Carboniferous Limestone, Millstone Grit and Coal Measures.

1 *Crescent, Buxton of Darley Dale stone.*
2 *Euston arch of Bramley Fall stone.*
3 *Williams and Glyn's Bank, Mosley Street, Manchester, of Wellfield stone.*
4 *Coventry Cathedral of Hollington stone.*
5 *Liverpool Anglican cathedral of Woolton stone.*
6 *Horsham stone on Sussex cottage roof.*

1

4

2

5

3

6

3.04 Carboniferous Limestone, predominantly a limestone in England, is found to contain an increasing number of sandstone beds as it is traced northwards through England into Scotland. This has helped make Northumberland, more than any other, a county of sandstone buildings. Prominent among Scottish stones, Craigleith stone has the same relationship to Edinburgh that granite has to Aberdeen.

3.05 The Millstone Grit division is probably the most prolific source of sandstones in the UK. In northern England it comprises a series of gritstones and sandstones interbedded with shales and mudstones. The Silsden Moor grit, Kinderscout grit and Rough Rock are the most important for building. Millstone Grit has been used for castles, cathedrals and churches from earliest times, **1, 2, 3**.

3.06 Despite the name, only about 5 per cent of the thickness of the Coal Measures division is coal. Many of the rocks are highly laminated and may be split into slabs 25 mm or less thick. They are generally known as York stone although the name is often used for sandstones from Yorkshire generally. The thin flaggy beds used for paving were quarried mainly around Sheffield, Halifax, Huddersfield and Barnsley.

Permian and Triassic—New Red Sandstone
3.07 These two systems are commonly grouped as New Red Sandstone as the stones they yield are similar and the allocation of some beds to a particular system is debated, **4, 5**.

Jurassic
3.08 This system has yielded a few sandstones, which were usually used locally. In Northamptonshire a division of the Jurassic known as the Inferior Oolite series contains the Northampton Sand Ironstones in its beds. Mostly worked for iron ore, some building stone has also been produced. These stones are responsible for the warm russet-brown hues of Northampton and many villages around.

Cretaceous
3.09 This system has yielded few building stones but some important ones can be found across southern England. Weald Clay contains calcareous sandstone beds which in some areas split readily. Known often as Horsham stone, these sandstones were originally deposited in shallow water and many slabs have ripple markings, **6**.

3.10 The Upper Greensand of the South Downs and Isle of Wight contains stone tinted by glauconite.

Tertiary
3.11 Few sandstones from this system are consolidated enough for building. Two oddities, sarsens and pudding stone, should be noted. Sarsens (or greywhethers) are hard massive boulders scattered on the surface of the chalk formation. Pudding stone contains pebbles and cementing matrix of equal hardness so that the stone shears through the pebbles, **7**.

Sandstone imports
3.12 The UK has a great variety of abundant sandstone. Though most other countries work sandstone, none has been imported for major building works.

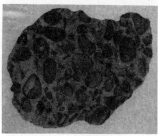

7 *Section of Hertfordshire Pudding stone.*

Table I Sources, characteristics and use of sandstones

Positions in geological column (age in million years)	Name of Stone	Source	Characteristics	Examples
Cambrian (500-570)		Caerbwdi quarry		St David's Cathedral, Pembroke
Devonian (345-395)	*Old Red Sandstone* (*Hereford stones*)			
	1 Red Wilderness stone	Forest of Dean, Glos.	Hard, reddish-brown, micaceous	Paving, steps, Liverpool Cathedral (Scott, 1908)
	2 Caithness flagstone	Caithness	Compact, tough, dark, blue-grey	Flags
Carboniferous (280-345)	*Carboniferous Limestone*			
	1 Prudham stone	Hexham, Northumberland	Coarse grained, slightly micaceous, brown	Newcastle Central Station. Additions to University College London. New office block above bus station, St Andrew's Square, Edinburgh
	2 Doddington stone	Wooler, Northumberland	Fine grained, pink to purple-grey	Dunblane Cathedral, Perthshire. Observatory and Wesleyan Hall, Edinburgh
	3 Blaxter stone	Elsdon, Tyne and Wear	Medium-grained, micaceous, buff	Many buildings in Northumberland
	4 Craigleith stone	Edinburgh	Fine grained, grey-brown	Many parts of Aberdeen
	5 Hailes stone	Edinburgh	Fine to medium grained, pink, white or drab coloured	Many parts of Edinburgh
	Millstone Grit			
	1 Darley Dale stone	Darley Dale, Matlock Derbyshire	Fine grained, compact, feldspathic, pale brown to white. Very strong	Embankment and Kings College Hospital, London Crescent, Buxton, **1**.
	2 Stancliffe stone	Darley Dale, Matlock, Derbyshire	Fine grained, compact, micaceous, drab coloured	St George's Hall, Liverpool.
	3 Hall Dale stone	Darley Dale, Matlock, Derbyshire	Fine grained, pink	City centre development, Lancaster
	4 Whatstandwell	Darley Dale, Matlock, Derbyshire	Coarser than 2 and 3, pinkish	Original Euston Station. Prisons at Leicester (1828) and Birmingham (1849)
	5 Birshover gritstone	Stanton Moor, Derbyshire	Medium grained, pink to yellow-buff	Public works in Lancashire. Bridges on M6. Royal Insurance building, Kirkcaldy
	(1–5 known as '*Dale Stones*') 6 Bramley Fall	NW of Leeds, Yorkshire	Coarse grain, pebbles up to 25 mm. Very strong, massive bedding	Engineering works. Euston Arch, **2**
	7 Spinkwell stone	Nr Bradford, Yorkshire	Fine grained, some mica, grey	Town Hall, Manchester
	8 Wellfield stone (a Crossland Hill stone)	Nr Huddersfield, Yorkshire	Fine grained, occasional quartz pebbles, brown-grey	Free Trade Hall, Manchester. New civic buildings, Newcastle. Williams and Glyn's bank, Mosley Street, Manchester, **3**
	9 Stainton stone	Barnard Castle, Co Durham	Fine to medium grained, buff	
	10 Dunhouse stone	Barnard Castle, Co Durham	Fine to medium grained, brown	Durham Castle and Cathedral. Sunderland Technical College
	Coal Measures 1 York Stones —Elland Edge stone	Brighouse, Yorkshire	Fine grained, micaceous, fissile	London pavements. Arches, Charing Cross
	—Elland Edge flagrock	Southowram, Halifax and Lane Head, Huddersfield		

Coal Measures continued	2 Bolton Wood stone	Bradford, Yorkshire	Fine grained, light greenish brown	Town Halls, Bradford and Leeds
	3 Woodkirk stone	Morley, Yorkshire	Fine grained, fawn to brown, massive	New Halifax Building Society hq
	4 Forest of Dean stone	Coleford, Glos.	Hard, durable, dark grey, micaceous and feldspathic sandstone, in places an arkose	New pier, Cardiff. New dock, Avonmouth
	5 Pennant stone	South Wales	As for Forest of Dean stone	Towns of most South Wales valleys. Bristol Jail. Worcester Cathedral
	6 Auchinlea stone	Cleland, Lanarkshire	Medium grained, white-cream, often flecked with brown	Edinburgh, Carlisle, Glasgow
Permian (225–280)	*New Red Sandstone* 1 Penrith Red	Penrith, Cumbria	Moderately coarse grained, bright red, pink to buff. Many grains have quartz outgrowths in optical continuity with grains	Much of Penrith
	2 Lazonby Red	Lazonby, Cumbria	Medium grained, red	Carlisle
	3 Locharbriggs stone	Dumfries	Fine to medium grained, red	Glasgow
	4 Mansfield stone	Mansfield, Nottinghamshire	Fine grained, white only (not red) now available, even grained. Quarried yellowish, it weathers white	King's College Cambridge (1823). Restoration Ely Cathedral
	5 Clashach and Green Brae stones	Hopeman, Morayshire	Even-grained, fawn	Old Course Hotel, St Andrews (Clashach)
Triassic (32)	*New Red Sandstone* 1 Hollington stone	Hollington, Staffs	Fine to medium grained, hardens on exposure, white, red, salmon and mottled	Hereford Cathedral (1148) Coventry Cathedral, **4**
	2 St Bees stone	St Bees, Cumbria	Fine grained, sometimes micaceous, bright red	Parts of Windsor Castle and Furness Abbey
	3 Grinshill stone	Shrewsbury, Shropshire	Fine grained, whitish grey (formerly red also)	Shrewsbury city, including Abbey Church
	4 Spynie and Rose Brae stones	Elgin, Morayshire	Fine grain cream or pinkish	
	5 Corsehill stone	Annan, Dumfriesshire	Fine grain, cream or pinkish	Synagogue, Great Portland Street, London. Town Hall, Reading. State Capitol, Albany, New York. Stockholm
	6 Woolton and Rainhill stone	Liverpool	Fine grained, red-orange	Liverpool Anglican Cathedral, **5**
Jurassic (136–190)	1 Duston and Moulton stones	Northampton	Buff or brown, with iron minerals	Northampton area
Cretaceous (65–136)	1 Sussex sandstone and Wealden Sussex stone	West Hoathly, Sussex	Fine grained, brown-yellow	Paving and roofing, **6**.
	2 Green Ventnor, St Boniface stone	Ventnor, Isle of Wight	Speckled light green to blue grey, fine grained	Structural use, Isle of Wight. Bell tower, Chichester Cathedral. Winchester Cathedral restoration (1825)
	3 Reigate stone	Reigate, Merstham, Gatton and Godstone, Surrey	Resists heat without breakdown in particle size. Massive compact beds (750 mm thick), greenish-grey with mica and much glauconite	Fireplaces, local building. Royal Palace of Westminster (1259). Windsor Castle. Henry VII Chapel at Westminster. Vaulting of crypt, Guildhall, London. (Reigate and Merstham stone)
Tertiary (2–65)	1 Satsens	South Downs	Quartzitic, fine grained, may contain rounded pebbles well cemented with silica	Stonehenge. Avebury Circle (600 sarsens). Windsor Castle. Churches in Surrey
	2 Pudding stone	Hertfordshire	Flint pebbles 50 mm diameter in matrix of sand and flint with silica as cement	Used for building in Essex (where stone is scarce) and millstones

4 Limestone

4.01 Limestones consist chiefly of calcium carbonate ($CaCO_3$), in the form of the mineral calcite and are formed usually either directly or indirectly from mineral matter dissolved in sea water. Limestones may be classified on the basis of their origin as chemical, organic and clastic (formed from rock fragments).

Chemical limestones
4.02 These are formed directly by precipitation of calcium carbonate from water. Circulating ground waters containing carbon dioxide, when passing through rocks containing calcium carbonate, will readily take it into solution. When the charged water reaches the surface the calcium carbonate is deposited as *tufa* or as *travertine*. In particular situations *stalactites* and *stalagmites* are formed. *Onyx marble*, discussed in detail later, is also included here.

4.03 Chemically precipitated limestones are also formed in sea water. Around the Bahamas, evaporation of warm shallow sea water in tidal channels and lagoons causes calcium carbonate to be precipitated in concentric layers, often round a fragment of shell or a grain of sand which acts as a nucleus. Such small, mainly spherical grains are known as ooliths. Some limestones composed mainly of ooliths are called *oolitic limestones* or *oolites*. The ooliths are commonly about the size of a pinhead, but in some rocks can be the size of a pea; the name *pisolite* is then used.

Organic limestones
4.04 These consist largely or entirely of fossilised shells. Spaces between shells are filled with broken shells or calcareous mud. Coral limestones in which complete coral colonies may occur and crinoidal limestones made up of crinoids ('sea-lilies', animals related to sea-urchins and starfishes), are included in this group. Many of the commercial 'marbles' are of this type.

Clastic limestones
4.05 These result from the erosion of pre-existing limestones, the fragments ranging from fine-grained calcareous mud to pebbles of the original limestone, later consolidated and normally cemented with calcareous material into a new coherent rock. Sutton stone from Glamorgan and 'Draycott marble' are of this type. Many commercial 'marbles' with the name 'breche' are clastic (detrital) limestones.

Mineral content
4.06 Few limestones are composed entirely of calcium carbonate; most contain clay which is one of the commonest non-calcareous constituents of limestones. The lower the amount of clay the better the polish which the limestone will take. Several other minerals may be present.

4.07 Considerable *dolomite*, the double carbonate of calcium and magnesium, is present in some limestones. The dolomite may have originated as a chemical deposit or by later alteration of the original limestone. Pure dolomite rocks are rare. The Carboniferous Limestone is dolomitised in patches, in some places on a large scale. Nearly all the limestones of Permian age in this country contain dolomite; one of the formations is known as the Magnesian Limestone, (referred to in TS 1). In general, dolomitic limestones are more resistant to weathering than 'pure' limestones.

4.08 Quartz, as sand grains, may be present to such an extent that the limestone is called a 'sandy limestone'. In some instances, complete removal of the calcium carbonate in a siliceous limestone leaves behind a 'skeleton' of silica. This is then known as 'rotten stone' and is used for polishing and as an abrasive.

4.09 Rarely, a limestone may be rich in one of the iron minerals. Some of the oolitic limestones of Jurassic age in Lincolnshire and in Northamptonshire are rich in iron minerals, commonly siderite (iron-carbonate) and chamosite (iron-silicate). The concentration of these iron minerals may be high enough for the stone to be worked as an iron-ore.

4.10 Glauconite is an occasional constituent of limestones but it is by no means as widespread in them as it is in sandstones (see **2.03**).

4.11 Pyrite (iron-disulphide) may be present and in disseminated form contributes to the dark colour of some limestones, some of which smell, when struck, of sulphuretted hydrogen.

5 Types of limestone

5.01 Limestones occur at most levels in the geological column. The text should be read with table II which classifies them, gives their common names, sources, characteristics and examples of their use and shows their position in the geological column. Limestone of Precambrian, Cambrian, Ordovician and Silurian age have not in general yielded useful building stone.

Devonian
5.02 The principal beds occur in a belt of country from Plymouth to Torquay. Several varieties of Devon 'marble' exist ranging from black to white, mostly veined or mottled with corals contributing to the *figuring*. They all weather to a pale grey.

Carboniferous
5.03 The Carboniferous system is the oldest in which limestone is a major constituent. It constitutes many areas of outstanding natural beauty; the Mendips, the Peak District and much of the Pennines. Many stones were taken from this system, cut, polished and sold as 'marbles'; many fewer are now in production[1] (see table II), **8**.

Permian
5.04 The Permian is made up of marls, sandstones, conglomerates and limestones. One of the limestones is known geologically as Magnesian Limestone, consisting of upper and lower limestones separated by beds of marl. It crops out in a strip rarely as much as 10 km wide from Nottingham through Mansfield and Tadcaster to Darlington where it widens to the coast at Hartlepool, Sunderland and South Shields. The building stones vary greatly from place to place, the best coming from the lower limestone.

Jurassic
5.05 This system has probably furnished more enduring and widely used stones than any other system. The deposits of this age are clays, shales, sands, sandstones (with some ironstones), and limestones. The limestones occur on a large scale and many are oolitic. So common are they that often this part of the column is divided into the *Lias* and the *Oolites*.

5.06 The Lias formation consists mainly of clays with thin-bedded limestones in the lower part which have been used locally for building. Limestones from beds following the Lias have had a greater effect on the general appearance

8

9

10

8 Derbyshire Fossil in Royal Festival Hall, London.
9 Knole House, Sevenoaks, Kent, a fine example of Kentish Ragstone.
10 Portland Roach, the Economist building, showing honeycombed appearance.

of more towns, especially in southern England, than any other stone.

5.07 Inferior Oolite (inferior refers to its position in the geological column), supplied vast amounts of the once widely used 'Cheltenham stone'. In Northampton, Lincolnshire and Rutland a series of beds called Lincolnshire Limestone form a large lens shaped deposit. The oolites from Lincolnshire are essentially similar rocks, though with individual characteristics.

5.08 The Great Oolite series is widely used, one rock being so important that its name is used geologically: 'Bath stone'. The series also contains fissile rock used as 'slates'. Limestones were available on a wide scale from so many quarries in the Cotswolds that they had the general name 'Cotswold stone'.

5.09 Another important oolite, Portland stone, is found in beds which form the tilted tableland of the Isle of Portland, the two arms of Lulworth Cove and the cliffs of the Isle of Purbeck. Warm cream in colour when quarried, it is so fine and even grained that the oolitic structure may be difficult to discern. A very pure limestone (95 per cent calcium carbonate) it weathers to a characteristic black and white colour in smoke-polluted atmospheres, the sheltered side being blackened.

5.10 The beds of Portland stone differ in character; building trade rather than precise geological names are used. A typical section through Portland stone beds is: (top) Roach, Whit Bed, Flinty Bed, Curf, Base Bed Roach, Base Bed or Best Bed (bottom). Each stone has individual characteristics, **10**.

5.11 The Purbeck beds contain numerous limestones individually distinguishable; Laning, Red Rag, Shingle, Spangle, etc, each of specific use; paving, building, tombstones, kerbs, setts, roofing. Of the Purbeck stones, Purbeck 'marble' is the best known.

Cretaceous
5.12 This system has yielded a few limestones, important especially in southern England. Weald Clay contains beds of shelly limestone which provide 'Wealden marbles', similar to Purbeck marbles. The Cretaceous takes its name from its major formation, the Chalk, a pure limestone (up to 98 per cent calcium carbonate). White to pale-grey in colour, very fine grained, uncemented and soft in southern England, it is fairly hard in northern England and has been used frequently for building.

Tertiary
5.13 Only one stone is important for buildings: Quarr stone from the Bembridge limestone formation. A mass of molluscs, it looks an unlikely building stone but was worked and used extensively during the middle ages and can be found in many Sussex churches.

Distinctive stones
5.14 Two stones, alabaster and flint, are not easily placed in any classification. When sea waters in New Red Sandstone times evaporated, gypsum was one of the first substances deposited. When this is in massive compact form, suitable for carving, it is called alabaster (and sometimes, wrongly, marble).

5.15 Flint, the origin of which is the subject of geological debate, is found as nodules, layers of nodules and more rarely as bands in the top of the middle and throughout the

Table II Sources, characteristics and uses of limestones

Position in geological column (age in million years)	Name of stone	Source	Characteristics	Examples
Devonian (345–395)	1 Devon marble	Ashburton (last quarry to close) Babbacombe, Petit Tor, Ipplepen, Ogwell	Grey to black, white and red patches and veins White, black, pink, dove, etc.	Steps and floors, Post Office Tower, London. Entrance of Geological Museum, London. Many shopfronts
Carboniferous (280–345)	1 Frosterly marble	River Wear, Durham	Grey to black with light coloured corals	Durham Cathedral (thirteenth century) Cathedrals at Peterborough, Bristol and Bombay
	2 Duke's Red	Rowsley, Derbyshire	Fine grain, deep red	St John's College Chapel, Cambridge
	3 Hopton's Wood stone (Hadene stone is indistinguishable) (1–3 are no longer quarried though stocks may be found in stone yards)	Hopton Wood, Derbyshire	Fine grain, cream	Geological Museum, V & A Museum and Imperial Institute, London
	4 Derbyshire Fossil	Coalhill, Derbyshire	Grey to dark fawn. When cut lighter crinoids are prominent	Royal Festival Hall, **8**, and Geological Museum, London
	5 Derbydene	Dene Quarry, Matlock, Derbyshire	As Derbyshire fossil	Thorn House, St Martin's Lane, London
	6 Orton Scar	Orton Scar, Cumbria	Light fawn with darker veins	
	7 Salterwath	Crosby Ravensworth Fell, Cumbria	Dark brown with feint markings	
	8 Deepdale fossil	Deepdale, Yorkshire	Light brown to dark grey with many fossil markings	
	9 Swaledale fossil	Barton, Yorkshire	Light brown with many fossils	
	10 Penmon limestone (Penmon 'marble')	Beaumaris, Anglesey	Light brown with darker veins and patches, blue-grey or grey-white, close grained, compact	Beaumaris Castle. Mersey docks
Permian (225–280)	1 Bolsover Moor	Bolsover Moor, Derbyshire	Warm yellowish brown, fine grained, dolomitic	
	2 Anston stone	Kiveton quarries, Sheffield	Light brown to cream, compact and fine grained	Restoration of flying buttresses, Westminster Abbey (1847). Record Office, Fetter Lane, London (1851)
	3 Other Magnesian Limestone: —Mansfield Woodhouse stone	Nottingham		
	—Steetly stone	Nottingham		
	—Roche Abbey stone	Yorkshire		
	—Huddlestone stone	Sherburn in Elmet, Yorkshire		Quarry reopened for repairs to York Minster
	—Tadcaster stone	To west of Tadcaster, Yorkshire		York Minster
Jurassic (136–190)	*Lias* 1 Blue lias	Station quarry, Charlton Macknell, Somerset		Ecclesiastical work such as detached shafts, bases and capitals, Bristol Cathedral
	2 Stowey stone	Stowey quarry, Bishop Sutton, Bristol		
	3 Hornton stone	Edgehill, Banbury, Oxon	Blue grey or brown to a greenish tint (sage green)	Local churches. Broughton Castle. Hilton Hotel, Stratford-upon-Avon
	4 Ham Hill	Hamdon Hill, Norton-sub-Hamdon, Somerset	Lenticular mass of fossil shell fragments, 15 m deep, worked for building stone. Richly toned brown	Montacute House, Somerset. Parts of St Anne's Cathedral Belfast. Hamstone House, Weybridge, Surrey

Age	Name of stone	Source	Characteristics	Examples
Jurassic (136–190)	*Inferior Oolite*			
	1 Guiting stone	Guiting, Glos	Warm brown to yellow (yellow Guiting) or white-cream (white Guiting), coarse with fossil matter	Used in north Cotswolds since Norman times. Stanway House (1630). Prinknash Abbey, Glos
	2 Doulting stone	Doulting, Shepton Mallet, Somerset	Light brown, coarse textured. Includes fragmented crinoids, cemented with calcite	Wells Cathedral, Interior of country offices, Taunton. Interior nave, Guildford Cathedral
	3 Collyweston 'slate'	Collyweston, Northants	Sandy, fissile, kept watered and exposed to frost, then 'clived'	Roofing 'tiles'. Guildhall roof, London. College roofs, Cambridge
	Lincolnshire Limestone			
	1 Ancaster stone (Freestone)	Grantham, Lincs	Varying shell content, cream-buff	Mansions and churches in Lincs. Cambridge colleges. Town Hall, Holborn, London
	2 Barnack stone	Stamford, Lincs (no longer available)	Shelly, coarse textured	Peterborough Cathedral. Ely Cathedral. Cambridge Colleges. Burghley House, Hunts
	3 Casterton stone	Stamford, Lincs	Coarser than most in the group, beige	Churches. Stamford buildings. Cambridge Colleges
	4 Clipsham stone	Clipsham, Oakham, Leics	Buff to cream, medium grained with shell fragments. Some blocks 'blue hearted', not a fault and weathers to greyish colour	External cladding, Berkeley Hotel, London. Restoration, Palace of Westminster
	5 Ketton stone	Stamford, Lincs	Yellow-buff, some blocks a distinct pink, beautifully regular, medium grained	Cambridge. City Hall, Norwich. Cotterbrook Hall, Northants
	6 Weldon stone	Corby, Northants	Pale brown to buff, fine even grained with shell matter and open texture. Frost resistant	St Paul's Cathedral (before fire). Kirby Hall, Northants. King's College Chapel, Caius, Jesus, Clare Hall and Sidney Sussex Colleges, Cambridge. Broughton House and Rushton Hall, Northants
	Great Oolite			
	1 Stonefield 'slates'	Stonesfield, Oxfordshire	Frosted, as Collyweston slate, (Inferior Oolite see above), a fissile sandy limestone, light brown	Oxford Colleges. Local roofing
	2 Cotswold 'slates'	Region from Andoversford through Naunton to Stow-in-the-Wold	As Stonesfield slates	Local roofing
	3 Bath stones —Stoke Ground —Winsley Ground —Bradford stone —Westwood Ground —Box Ground stone (St Aldhelm stone) —Combe Down stone —Monks Park stone (Westwood Ground and Monks Park stone are still worked)	Bath, Somerset Bradford on Avon, Wiltshire Corsham, Wiltshire	Pale brown to light cream, even grained, poorly fossiliferous Paler and finer grained than others	Apsley House, Hyde Park Corner North and west faces of Buckingham Palace Sun Alliance House, Bristol
	4 Tayton stone	Taynton, Oxfordshire	Brown with light streaks, coarse and shelly	Early Oxford Colleges (Magdalen, Merton, St John's). Interior of St Paul's
	Portland stone			
	1 Roach		Cream, honeycombed by empty moulds of fossils. In some moulds shell casts may remain	*Economist* building, St James's, London, **10**. Commercial Union House, Birmingham
	2 Whit bed and Base bed (indistinguishable in use)			Many London buildings; Britannic House, Bush House, County Hall, Shell Centre, Bank of England. Civic Hall, Leeds. Reference Library, Manchester. Municipal Buildings, Southampton. Bank of England, Birmingham
	3 Purbeck-Portland		Beds 35 m thick	Many churches (using Pond and Under Freestone)
	(1–3 are types, not stones from individual quarries)			
	4 Chilmark stone	Chilmark, Vale of Wardour, Wilts	Sandy, glauconitic, greenish-grey	Wilton House, Wilts. In Salisbury Cathedral
	5 Tisbury stone	Tisbury, Vale of Wardour, Wilts	Sandy, glauconitic, greenish-grey. Difficult to distinguish from Chilmark	Winchester College
	6 Purbeck 'marble'	Isle of Purbeck	Mainly a mass of fossilised shells, bands (in ground) only up to 300 mm thick, takes high polish, blue, green, brown, red	Some shafts of Galilee Chapel, Durham Cathedral. Used 'in bed' in Ely Cathedral
Cretaceous (65–136)	*Weald Clay*			
	1 Large Paludina marble (Sussex marble) —Bethersden marble —Petworth marble —Laughton stone	 Bethersden, Kent Petworth, Sussex Laughton, Sussex	Shelly, similar to Purbeck marble	Altar steps, Canterbury Cathedral
	2 Small Paludina marble —Charlwood stone	Charlwood, Surrey	Easily confused with Purbeck marble	
	Lower Greensand			
	1 Kentish ragstone	Kent area	600 mm beds alternating with sand called 'hassock', dark blue to green-grey, sandy, glauconite, tough, intractable	Many churches. Walls of Londinium. Knole House, Kent. City Prison, Holloway (1849–52) **9**.
	Chalk			
	1 Amberley stone	Amberley, Sussex	Large blocks, white to pale grey, fine grained	Vaulted ceilings, Arundel Castle. Chichester Cathedral
	2 Beer stone	Beer, Devon	White to pale grey, mainly fairly coarse shell fragments	St Stephen's crypt, Houses of Parliament. Exeter Cathedral. Carved interior, Cathedral at St Louis, Missouri, USA
	3 Totternhoe stone	Dunstable, Bedfordshire	Shell fragments give gritty feeling (often incorrectly thought of as sandy), greenish grey	Woburn Abbey (quarrying resumed for stone for repair)
	4 Burwell stone *(3 and 4 commonly called 'clunch')*	Burwell, Cambridgeshire	Similar to Totternhoe	Arcading of Lady Chapel, Ely Cathedral
Tertiary (2–65)	1 Quarr stone	Near Quarr Abbey, Isle of Wight	Mass of molluscs, normally seen as casts or moulds, shell sometimes replaced by calcite	Many Sussex churches. In Winchester Cathedral, Chichester Cathedral and Lewes Priory
Distinctive stones	Alabaster	Fauld, Burton-on-Trent and Chellaston, Derbyshire	Massive, compact, suitable for carving	Tomb effigies. The 'Marble hall' of Holkham Hall, Norfolk
	Flint	Kent, Surrey, Sussex, East Anglia	Primary flint is black or dark grey. Can be split or 'knapped' to give a flat surface	Many buildings, especially churches in Kent, Surrey, Sussex, East Anglia and London. Guildhall, Norwich shows the celebrated chequer work using flint and Caen stone (for Caen stone see 6.02)

uppermost divisions of the Chalk formation. It is a compact cryptocrystalline (consisting of microscopic sized crystals) silica and breaks with a 'conchoidal' fracture leaving a shape and sharp cutting edge valued by early man for flint tools.

6 Imported limestones

6.01 There is a wealth of limestone in UK used from an early date. Most imports have been to use limestones, cut and polished as 'marbles', for decorative effect. Colour is the principal factor controlling choice. This text should be read with table III which classifies them by country of origin, name, source, characteristics and examples of their use.

6.02 France produces both true marbles (*marbres*) and limestone (*pierre marbières*); over 200 French 'marbles' are known. Most limestones are light coloured, those from the Ardennes next to the Belgian frontier are virtually identical with Belgian stones. The limestones are Devonian, Carboniferous, Jurassic and younger in age.

Limestones imported for repair work to buildings of Caen stone, itself the oldest import, are poor matches.

6.03 Belgium produces Carboniferous age 'marbles' and is one of the world's leading marble producing countries. Much is also imported into Belgium, processed and exported.

6.04 Israel actively exports. Some limestones have been used decoratively since biblical times.

6.05 Italy is renowned for marble, especially Carrara, though many limestones are also quarried. Travertine is one of the best known, a chemically precipitated limestone deposited from cold streams or lakes.

6.06 Portugal has lately developed production of decorative stone. Both true marbles and limestones of Jurassic age are worked and sold under a plethora of names. Much of the material imported to the UK comes via Belgium. Spain also has a massive output of 'marbles', some of them limestones.

6.07 Most countries produce 'marbles' and many are now major producers. A few can be found in the UK.

6.08 Confusion exists in the use and misuse of the names onyx and onyx-marble. Onyx is a minutely crystalline structured variety of quartz, banded black and white. It cannot be scratched with a knife. Onyx-marble is an exceptionally fine grained, generally translucent variety of calcite, considered to have been chemically deposited from standing sheets of water. It can be scratched with a knife.

6.09 Onyx-marble, normally banded, ranges from nearly white, ivory, yellow, green, reddish, red-brown to brown. It is prized for its translucence and as a luxury material—for marble bathrooms. It is now produced in many countries: Italy, Argentina (Brazilian onyx), Algeria (possibly the largest producer), Turkey, Iran, and Pakistan are very well known. Mexican onyx-marble, used by the Aztecs is widely used; for instance in the entrance of the Trocadero restaurant, London (now demolished). A fantastic use of onyx-marble from Turkey can be seen in the Piccadilly Circus branch of Barclays Bank, **11**.

11 *Banking hall clad with onyx-marble.*

Reference

1 For a complete list of producing quarries see *Natural Stone Directory* (see page 6).

Table III Sources, characteristics and uses of imported limestones				
Country of origin	Name of stone	Source	Characteristics	Examples
France	1 Caen stone	Caen	Fine grained, yellowish or yellow-white	Many ecclesiastical uses. Canterbury Cathedral
	2 St Maximin —Moulin a vent —Richemont —Lepine —Longchant stone	Oise Meuse Char/Marit Vienne	Jurassic, light buff, fine grained. Longchant is markedly oolitic	
Belgium	1 Rouge royal —Rouge Griotte —St Anne —Bleu belge —Petit granit (sold as marble, in fact a limestone)			
Israel	1 Various 'marbles'			Some Marks & Spencer shop fronts. Banco di Roma, Brussels
Italy	1 Verona	Venezia and Lombardy	Jurassic age limestone, including red and yellow Verona	
	2 Aurisina and Nabresina (Roman stone)	Trieste and Istria (partly now in Yugoslavia)	Cretaceous and Tertiary limestones including Aurisina and Nabresina. Nabresina resembles Hopton Wood stone and is an accepted substitute.	
	3 Verde Fraya	Piedmont	So-called modern 'green-marbles' are from the Piedmont and are mostly serpentinite, such as Verde Fraya	
	4 Siena marbles and Travertine (also USA)	Tuscany (travertine in Siena and Tivoli districts mostly)	Yellow limestone. The Siena 'marbles' are very important. Travertine may be loose, friable, often very porous. It ranges from cream through yellow to dark brown	Flooring, walling and cladding. London Transport Building, St James's. Cloaca maxima, Rome
Spain	1 Brocatello		Fossiliferous limestone varying widely in colour	Used in Westminster Cathedral (along with most other decorative stones)

Geology Technical Study 4

Metamorphic rocks

This study by FRANCIS G. DIMES describes the varieties of metamorphic rocks used for building purposes. He explains their geological form and gives examples of their use.

Formation
1.01 Metamorphic rocks are formed by the crystallisation or recrystallisation of pre-existing rocks at elevated temperatures and pressures under the earth's surface. The original constituent materials of the rocks (which may have been formed under widely different conditions) are rearranged mechanically or chemically, usually with the development of new minerals. During metamorphism the original rocks lose many if not all of their original characteristics.
1.02 Metamorphic rocks exist in large number. The common rocks are schist, quartzite, gneiss, serpentinite, marble and slate.

Schist
1.03 One of the most abundant of the metamorphic rocks, it is a laminated (foliated) rock built up from mica flakes, 2. Many varieties exist characterised by the presence of distinctive accessory minerals such as garnet, kyanite, chlorite and biotite. Because of the discontinuous and uneven thickness of the lamination or foliation, schist has been rarely used for building purposes. However, many quartzites of commerce are schists. These differ from ordinary quartzites simply in having a laminated texture produced by planes of mica flakes.

Quartzite
1.04 If a sandstone is metamorphosed the constituent grains are recrystallised to an interlocking mass. Mica flakes and occasionally other minerals may be present. (It should be noted that if the quartz grains of a sandstone are cemented by secondary silica, the rock is also called a quartzite. The term is descriptive; it is not an indication of the mode of origin.) Hard and strong, quartzites are finding increasing use for flooring and cladding.

Gneiss
1.05 A coarse-textured and banded rock made up of alternating layers of dark mica-schist and pale granitic (quartzo-feldspathic) material, 3. Some gneissic rocks are commercially produced as granite.

Serpentinite
1.06 Composed mainly of the mineral serpentine, the rock normally is green in colour, but may have red colouration. Formed by metamorphism of basic and ultrabasic rocks, it is usually found in assocations amongst highly folded schists of major mountain ranges. Probably it was formed from slivers of former ocean floors that escaped being carried down subduction zones and adsorbed in the earth's mantle when moving plates of the earth's surface collided. The serpentinite of the Lizard, Cornwall appears to be an exception (see p. 1). Genoa Green is an example of a serpentinite.

Marble
1.07 Geologically the name marble is restricted to limestones that have been completely recrystallised by heat or pressure. The calcium carbonate ($CaCO_3$) of the limestone is converted into crystals of calcite. A pure limestone is converted into a pure white marble such as the 'first statuary' marble of Carrara, Italy. If the original limestone contained sand, clay and other 'impurities', and recrystallised in the presence of water, serpentine may be formed leading to a green, banded or figured stone. It is an important constituent in many of the commercial 'green-marbles'. Connemara marble may be instanced.

Slate
1.08 When clayey rocks are subjected over a period to pressure a new grain is imposed on the rock. The mineral constituents are closely packed, may be flattened and are partly reconstituted to give new minerals, mainly mica, which lie at right angles to the pressure. The rock acquires a slaty cleavage which may be of variable thickness, 4. The original bedding of the clay may sometimes be seen as coloured bands running at a different angle to the cleavage surface.

1 Mountain side of bare marble showing rough schistosity. Mount Jagro, Carrara, Italy.

2 Schist; closely laminated layers of mica.

3 Gneiss; coarse mica with garnet (dark) and granite (quartzo-feldspathic) bands.

2 Stones for building

2.01 Stones in this study are classified by rock type. The text should be read with table I which classifies these groups by source, name and characteristics and gives examples of their use[1].

Table I Name, source and characteristics of metamorphic rocks

Metamorphic type	Name of stone	Source	Characteristics	Examples
Schist	1 Alta quartzite (Altazite)	Alta, Norway	Flaggy, mica-quartz-schist, 85 per cent mineral quartz. Grey.	Flooring of Bull Ring shopping arcade, Birmingham. Floor and column cladding, St Vincent House, London. The Suede Shop, Edinburgh. Paving of Royal Courts of Justice, Strand, London
	2 Otta Slate (Pillaguri slate or Rembrandt stone)	Pillaguri mountains, between Oslo and Trondheim, Norway	A garnet-hornblende-biotite-muscovite-quartz-schist. Blue-black. Hornblende may appear as needle-like streaks up to 50 mm long; garnets as small red spots. When polished has 3-dimensional appearance. Some surfaces are golden to copper-brown rust	Hertie Shopping Centre, Bonn, Germany. Bergen Hospital, Norway. Unilever House, Rotterdam. William Hill Organisation, Brompton Road and Nationwide Building Society, High Holborn, London
	3 Barge quartzite, (Sanfront stone, Italian quartzite)	Mount Bracco, Italy	Mica-quartz schist with up to 95 per cent quartz. Grey, gold, amber, olive	Bear and Staff pub cladding, Charing Cross Road, London. Facade, Radiant House, Liverpool Gas Co. Shell Centre, London, 5
Quartzite	1 Diamantzite	Namaqualand, South Africa	Tough, mica along marked planes giving very smooth cleavage, white, off white, grey and pinkish	Paving of Post Office's Monarch telephone exchange, London. Nationwide Building Society, Edinburgh
	2 Safari quartzite	Transvaal, South Africa	Silver-grey to green, similar to Diamantzite	
Gneiss	1 Alps grey 'granite'	Switzerland	A biotite gneiss, white and dark grey to black	
	2 Ematita 'granite' (Verde Ematita or Madreperla)	Uruguay	Formed by contact metamorphism (baking by molten igneous rocks). Blue-grey to greenish	Marks & Spencer shops in Northampton, Sheffield, Cardiff and Oxford Street (eastern end) London
Slate	1 Ballachulish slates (not now in production)	Ballachulish, Argyll	Blue-grey to black, some had pyrite crystals ('diamonds'). At one time also mottled green and purple slate called 'tartan'	
	2 Westmorland green —Broughton Moor —Spoutcrag —Elterwater —Kirkstone green —Lakeland green —Buttermere	Lake District and parts of Lancashire & Cumbria	Various shades of green; olive, light sea, pale barred, etc	Mullard House, Tottenham Court Road, London. Hotel Leofric, Coventry. ICI research labs, Alderley Edge, Cheshire. Bank of New Zealand, Christchurch. Imperial Bank of Commerce, Imperial Square, Montreal, Canada, where 1·5 ha of slate cladding was used
	3 Burlington slate	Kirkby-in-Furness, Cumbria	Metamorphosed during Caledonian orogeny, black to blue-black or blue-grey	Cladding of John Dalton House, Manchester and civic centre, Taunton
	4 Brathay slate	Ambleside, Cumbria	As Burlington. May show small amounts of brassy yellow mineral, pyrite. Worked since Elizabeth I, recently reopened	Standard Bank, St James's House, Manchester. Flooring of Civic Centre, Doncaster
	5 Welsh slates	See reference 2		Floor of St Bartholomew's Church, St Albans. Cladding of tunnels on Newport to Monmouth road. At Montparnasse station, Paris
	6 Delabole slate	Delabole, Cornwall		V & A Museum, London. Truro Cathedral
Marble	*British Marbles* 1 Iona marble	Isle of Iona	White with green to yellowish-green streaks, bands and mottling	Old altar, Iona Cathedral. Pavement of St Andrew's Chapel, Westminster Cathedral
	2 Skye marble	Torrin quarry, Isle of Skye	Previously used for statuary, now produces white crushed stone and aggregate	
	3 Connemara marble	County Galway, Eire	Greyish groundmass, in which light to dark green serpentine occurs as twisted and interlocking bands. In block it ranges from white to dark green	Ornamental arch over staircase, Geological Museum, London. Pillars in St John's College Chapel, Cambridge. Presbytery steps, Worcester Cathedral. Chancel floor, Peterborough Cathedral
	Imported marbles 1 Australian	Australia		Australia House, London
	2 Campan —Campan vert —Campan mélange —Campan rose	Campen valley, Hautes Pyrénées, France	White coloured, elongated almond-shaped masses, closely packed, cemented with pinkish or greenish matrix. The stone was not completely metamorphosed	Pavement tables of Paris cafes. Sacred Heart Chapel, Brompton Oratory, London
	3 Jaune lamartine	Jura, France	Yellow to yellowish-brown with irregular red and violet veins	Many J. Lyons restaurants
	4 Parian marble	Isle of Paros, Greece	Considered the finest white marble in the world. In places a sparkling play of colours	Parthenon roof, Athens. Temple of Apollo, Delphi, (530 BC). Drapers' Hall, London
	5 Pentelic marble	Greece	Rivals Parian, white and blue (or Greek Dove), some with cloudy markings	Propylae, Pericles, (437 BC) and many Greek buildings. Euston Tower, London. Interior walls and floors, National Westminster Bank, King Street, Manchester, 6 Strand Palace Hotel, London. Canadian Pacific, Cockspur Street, London. Altar, Peterborough Cathedral
	6 Skyros marble	Isle of Skyros, Greece	White with orange to golden-yellow to brown or violet veins and markings. Varied, specimens look widely dissimilar	
	7 Verde Antico	Larissa, Greece	Most prominent of 'green marbles'. Brecciated dark green serpentinite with white coloured calcite masses	Columns of St Sophia, Istanbul, Turkey. Used extensively for UK ecclesiastical work and present day shopfronts
	8 Tinos	Isle of Tinos, Greece	Also a serpentinous marble, mostly dark green with lighter green veins, with little calcite	Many shopfronts
	9 Carrara marble —Bardiglio —Arabescato —Pavonazzo —Cipollino	Carrara, Italy	—Bluish-grey —White with grey markings —Dark purple veins and markings —Banded white and green	
	10 Verde Fraye	Piedmont, Italy	Serpentinous marble, dark green with lighter streaks, veins and sometimes patches and veins of green or white stained calcite	Shopfronts
	11 Breche rose (Norwegian rose)	Norway	Brecciated, white and pink	Many decorative uses. Altar of Chapel of St Augustine, Westminster Cathedral
	12 Swedish green	Sweden	Greyish-white with twisted bands, streaks and patches of green to brown	Chapel floor, Manchester Cathedral. Concourse, Scottish Life House, Bridge Street, Manchester. Coliseum Theatre, London

Schist, quartzite, gneiss

2.02 Schist, quartzite and gneiss are produced in a variety of colours, polished, some with elaborate mineral effects showing on the surface. Used decoratively for claddings and pavings, they are all imported, **5**.

Slate

2.03 Slate has found wide use for roofing, stair treads, flooring, table tops and acid tanks because it is chemically inert, impervious and can be split (cleaved) smoothly and readily into thin or thick sheets. In UK, slate is found chiefly in Scotland, the Lake District, north Wales and Cornwall.

2.04 At one time Wales produced more slate than all other areas put together; the prime producing areas were around Corwen, Blaenau Ffestiniog and Corris and along a belt of country from Nantlle to Bethesda. Some was also produced at Mynydd Presely, Pembrokeshire, south Wales. Lately several quarries have closed but Welsh slate is still available in quantity and a variety of colours, deep blue, blue, dark grey, blue-grey, green, bronze and 'heather'. Used for roofing, once the predominant material in London, it has found a variety of other uses and increasingly is being used for cladding. The Lake District now is a major producing area. The slates, derived from a volcanic ash mud, are commonly known as 'Westmorland Green'. They have been widely used for roofing and for cladding.

2.05 Slate has been imported, 'cheaper but also inferior',[2] from France and Belgium and lesser amounts from Spain and Portugal. Italy, though associated with marble, works slate, such as at Chiavari. There is a large output of this in the forms of sills, 'tiles', skirtings plus 2000 to 3000 billiard table beds per month.

Marble

2.06 Few countries do not produce marble, although many 'marbles' sold are in fact limestones. Even so the range is enormous. The choice for decorative use, as distinct from use as sills, cladding and other building purposes, is on colour. Panels of highly coloured marble can be seen hung on walls, as with paintings.

2.07 In the British Isles marbles are rare rocks and few of them can be used for building. Those most widely used are from the isles of Tiree, Iona and Skye and from Connemara, Eire.

2.08 Many countries produce marble imported over the years into the British isles. France is a major source, working marbles since Roman times. Most are coloured, there is little white; the classic stone is Campan. Greece produces the classic white marbles Parian and Pentelic, the highly coloured Skyros and the serpentinous marbles such as Tinos and Verde Antico, **6**.

2.09 Italy claims first place in the production of marble. Carrara is the best known for pure white (and to some characterless and boring) marbles.

Carrara and the surrounding area is the largest marble producing region in the world. The marbles may be grouped:

   i Bianco Chiaro (or Blanc Clair); white with a few greyish markings.

   ii Bianco Chiaro Venato (or Bianco Venato or Blanc Veine); white with stronger greyish markings which may adopt a vein pattern.

   iii Statuario (or Statuary Marble); divided into **a** First Statuary, virtually pure white, **b** Second Statuary, with some grey markings, **c** Vein Statuary.

   Commonly it is difficult to distinguish between **b** and **c**.

   iv Bardiglio (or Bleu Tarquin or Italian Dove); 'bluish' grey. The colour varies and the different markings lead to many names being given to it.

Coloured marbles also are produced, notably, Arabescato, white with grey markings; Pavonazzo, with dark purple veins and markings; and Cipollino, banded white and green. Many Italian 'marbles' were considered in the limestone section of TS3.

2.10 Norway has a lot of marble though little is exploited, Breche Rose being the best known. Similarly Sweden works little of its marble: the best known in the UK is Swedish green, which has been quarried since 1650. Vast quantities of marble also exist in the USA but none is currently imported into the UK.

Serpentinite

2.11 Verde Fraye is typical of the modern 'green marbles' of the Piedmont, Italy. It is dark green with lighter green streaks and veins. Green 'marbles' of this area are marketed under dozens of names (all, however, described as 'Verde').

4 Slate quarry, Croyton, Tavistock, Devon: well-developed cleavage planes.
5 Italian schist cladding to Shell Centre, London.
6 Pentelic marble from Greece in Natwest bank, King Street, Manchester.

7 *Verde Fraye, a serpentinite from Italy, facing shop front in Bond Street, London.*

They have been extensively used for shop fronts and other decorative work.

3 Geological Museum

3.01 No complete collection of stones suitable for building exists. However the Geological Museum (Exhibition Road, London SW7) has an exhibited series of stones backed by an extensive reserve and reference collection. These include representative building stones from most quarries now working in the UK and from many disused ones which can be an aid to matching stone for repair. There is also an extensive series of marbles and other ornamental stones from most countries of the world. These collections can be inspected by appointment.

References

1 For a complete list of producing quarries see *Natural Stone Directory* 1977 Ealing Publications, 73a High Street, Maidenhead, Berkshire.
2 F. J. North, *The Slates of Wales*, National Museum of Wales, Cardiff, 1946.

Acknowledgements

I should like to thank Martyn Owen and F. W. Dunning (curator) of the Geological Museum for their helpful comment. These four technical studies are published by permission of the Director, Institute of Geological Sciences, Exhibition Road, London SW7 2DE.

Quarrying Technical study 1

Nature and location of quarries

Architects should not only see stone used in building but visit quarries to see how the material is won and worked and its quality and colour. PHILIP BURTON introduces the current pattern of quarrying, referring to imports as well as UK production.

1 Introduction

1.01 Visiting quarries is a tradition stretching back long before Wren selected blocks of limestone at Portland, and it is one that is followed today (though not often enough) in various granite, marble and slate quarries in most countries. Because of the way it has been formed, a rock may vary in pattern and colour quite considerably, though this will not be evident from a hand-sample in a masonary showroom.

2 Stone quarries in Britain

2.01 There are about 200 British quarries now producing almost that number of block limestones, sandstones, granites and slates. Some of these are identical in colour and texture with materials produced centuries ago and used on buildings which are still standing—very fortunate when restoration or repairs are needed. A century ago it would have been possible to identify two to three times the number of sources left today. While the specifier has been deprived of some useful stones, there is still an outstanding variety available suitable for building.

2.02 The decline in the number of quarries has slowed in the past two decades. The more recent closures have been caused by many factors: beds have been worked out or have become inaccessible, planning restrictions, lack of capital for proper development, or, less often, shrinking demand. These closures have meant the loss of some well-known materials even in the last 10 years.

2.03 The still-thriving Scottish granite trade has lost the Rubislaw grey and Creetown white, and has only the Peterhead in production. The north Wales slate industry which once offered roofing slates from a choice of 30 to 40 quarries is now down to a half-dozen, losing Dinorwic, Dorothea, Oakeley and a multitude of others. In Cornwall 20 years ago, more than 30 quarries could be named, compared with a few today. Among the decorative limestones, output from Bethersden, Petworth, Hopton Wood and Ashburton has long ceased, though small quantities of the latter two can still be tracked down in masonry yards, and other varieties have become available.

2.04 Despite this the stone industry has remained active and adapted itself to current needs. Stone production now relies as much on machines as craftsmanship, and efficient sawing and polishing equipment is needed in the workshops, while modern earthmoving and other plant is often necessary at the quarry face. A diamond-bladed frame saw costing in the region of £20 000 is a typical investment even for a single machine, and is rarely possible for very small firms, **1**.

1 *Frame saw; cuts Portland stone at 425 mm per hour, marble at 300 mm.*

Quarry ownership

2.05 Quarry ownership has become more varied though many quarries have links with other branches of the construction industry. The biggest participant is probably the Bath & Portland Group whose Kingston Minerals Ltd operates the principal quarries for Portland, Bath (Monks Park and the re-opened Westwood Ground), Doulting and Guiting limestones. The De Lank quarry of the Thos W. Ward group is a principal source of Cornish granite (Tarmac's Hantergantick is another) and this company also quarries the pink Shap granite in Cumbria and the Ketton limestone of Lincolnshire. Sir Alfred McAlpine & Son Ltd has a major interest in the big slate deposits of Penrhyn Quarries Ltd near Bangor. Others are owned by companies with interests in crushed material, by plant hire firms, builders, stonemasonry companies, family firms, and a substantial number of individuals. Nearly all are run by managers and men who have spent their lives in stone quarrying—and usually with a particular type of stone.

2.06 The gradual grouping of some quarries under common control has offered the benefits of pooling both marketing and technical resources. Among these groups are Bolehill Quarries Ltd, Natural Stone Quarries Ltd, Johnson Wellfield Quarries Ltd, Cumbria Stone Quarries Ltd and Hard York Quarries Ltd. Lake District slate producers who have created links in the past year or so are the Lakeland Green and Burlington quarries, and Broughton Moor with Delabole in Cornwall. For contracts beyond the capacity of an individual quarry, the industry can pool resources when the need arises, as was shown when York Stone Firms Ltd was set up several years ago as a consortium by independent quarries.

Stoneworking facilities

2.07 The skills and knowledge of the quarryman, although vital in producing good stone, are now complementary to those of the stone machinist **2**. The simplest workshop may consist of little more than a guillotine, but more advanced, modern yards attached to quarries have circular, frame or wire saws **3**, as well as equipment for polishing, milling and surface texturing **4**.

2.08 The availability of this technology has widened the range of material which many quarries can now offer. At one time for instance the high-quality slate of north Wales and the Lake District was produced and riven solely as roofing slates, and these were sold all over the world. However, fast cutting and polishing equipment now enables the slate to be worked as cladding slabs, steps and risers, sills, and for many other purposes. This has created a demand for large blocks which can be won from the quarry face with a minimum of waste.

3 Imported material

3.01 Because of Britain's variety of limestones, sandstones and slate, relatively small quantities of these are imported. But there is a long-established import trade for the great variety of colours and patterns of granites and marbles which are not to be found in our own quarries but which are needed for a high proportion of decorative work on and in hotels, banks, insurance premises, airports, shopfronts, etc.

3.02 Italy is the focus of the world marble industry, for not only does it have about 2000 quarries of its own, but its various sawing and polishing works claim to handle as much as 80 per cent of the world output of marble. Many varieties of marble are imported to Britain from Carrara, Verona, Rome and the other centres, but there is also a traditional import trade from Spain, Portugal, Yugoslavia, North Africa, Belgium, France, Greece and Switzerland (see also table I).

3.03 Among igneous stone imports the main sources supplying Britain have been Finland (Balmoral Red), Sweden (Bonaccord Black, Ebony Black, Imperial Red, etc), Norway (Blue Pearl, Emerald Pearl), South Africa (Pretoria Black) and Brazil (Andes Black). Much of the granite worked in Britain comes through Aberdeen where there is still the greatest concentration of modern cutting and working machinery.

3.04 There are approximately 100 firms in Britain which regularly import granites and marbles[1]. Much of this material is brought in as blocks and is subsequently cut and polished for use either by the importers themselves if they are masonry contractors or, if they are wholesalers, by other companies in the trade who may be responsible for cutting fixing-holes and other finishing work. Material is often imported in slab form by the container load to reduce transport costs of waste.

4 Future of the quarries

4.01 If many of our natural resources have a predictable end, this is not yet in sight for most building stones. Slate has been won from the Lake District, for example, for several centuries, and apparently can be quarried there for at least as long again into the future. And, say the quarrymen, the quality gets better the deeper they go. And looking at the mountains outside the town of Carrara, there is little prospect of a shortage of what is generally regarded as the best marble in the world. These, like most of the materials still worked and won by the stone trade, were used long before the Romans employed them and will be available for several centuries more.

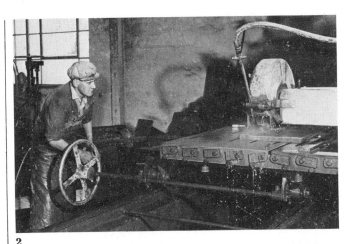

Table I Main sources of imported materials used in Britain

Granite	Marble	Quartzite	Onyx-marble	Limestone	Slate
Sweden	Italy	Italy	Argentine	France	France
Norway	Greece	South Africa	Pakistan	India	Spain
Finland	Spain	Norway	Mexico		
Ireland	Belgium	Sweden	Turkey		
Brazil	France		Algeria		
South Africa	Ireland				
Portugal	Yugoslavia				
Angola	Switzerland				
India	Morocco				
USA	Sweden				

2 Cross cut saw cutting Burlington slate.
3 Wire saw; wire copiously fed with water and abrasive cutting Burlington slate.
4 Bush hammering Cornish granite.

Reference
1 For a list of UK quarries, importers and agents see Natural Stone Directory 1977, see page 6.

Quarrying Technical study 2

Quarrying methods

The architect should have some knowledge of the workings of a quarry to be able to deal with specialists and to make use of their advice. JOHN ASHURST and ARTHUR FORDHAM* describe quarrying methods, including notes on specification to the quarry.

1 Introduction

1.01 If architects are not experienced enough to select their own stone from the quarry face they must make use of an inspector or rely on the supplier. But even the most inexperienced should be able to recognise the presence of the more serious natural defects which sometimes occur in stone, such as the presence of soft beds or small vents, and should know the state of a quarry well enough not to specify impossible sizes unrelated to the bed formation. They should be familiar with the likely colour range and the character and variety of texture which may be available. They should be assured that there is sufficient stone running in beds of sufficient height to complete the contract, and should take advice early enough on detailing to ensure maximum speed and economy in production. These notes on extraction and processing give the necessary background.
1.02 Developments in mechanical extraction of stones from the quarry or mine source are tending to reduce traditional differences in quarrying methods between different stones. Thus the same or similar methods may now be used both on slate and granite, or on limestone and sandstone. Although change from old and wasteful systems of extraction is too often very slow, mechanisation is spreading.

2 Limestone and sandstone

Use of 'waste'
2.01 Beds of good building stone may occur near the surface, sometimes within 2-3 m as with the St Bees or Darney sandstones, or they may occur 20 m or more below, as with some of the Portland freestone. The cost of removing the overburden is of course always relative to the use to which the overburden may be put. The upper strata may contain tough, well broken ragstone, suitable for dry walling. More usually, it will be processed for use as roadstone, hardcore and free draining material. At Portland, the rubble removed to expose the freestone is backfilled into discontinued workings, and the 'top cap' is reduced to sizes suitable for manufacturing high quality concrete units.
2.02 Limestone is also crushed for roadstone and for use in steel production. Other markets include specialised ciment fondus, agricultural lime and concrete blocks. Sometimes, as at Ketton, the freestone is a subsidiary product of the quarrying operations, which are principally feeding a cement works.

Extraction
2.03 Frequently the sedimentary freestones[1] lie in well defined near-horizontal beds varying in height from small, well laminated material a few mm thick to more massive strata 2 m thick. Frequently it is possible to work the stone entirely to its bedding planes and joints without having to split away from the solid. Some of the naturally occurring fissures and fractures are quite small, but others are wide enough to step into, and are reasonably consistent and parallel to each other, **1, 2**.

[1] A 'freestone' may be defined as a stone which can be worked freely in any direction.
* Based on material supplied by J. V. Borthwick, J. Reay, K. Jennings, J. M. Reid, T. Germain.

1 *Limestone quarry face showing natural bedding and jointing.*

2 *Freed sandstone block being lifted from bed.*

3 *Feathers (split rods) are dropped into drilled holes and plugs (wedges) driven in after to split stone.*

4 *Mechanisation in the quarry. Mobile chain and wire sawing.*

5 *Splitting with plugs and feathers.*

2.04 The equipment and the methods of extraction have changed very little over the past few hundred years. Equipment consists of kivels, twibills, crowbars, sledge hammers, plugs and feathers, and drills. The main developments in recent years have been the introduction of mechanical equipment for the removal of overburden, the use of electrically operated derricks in place of steam or hand operated varieties, and compressed air drilling.

2.05 Where no natural joint is available, a drilled section is prepared and split. A row of holes about 9in apart is drilled along the line of the desired split, and the block is then wedged off using 'plugs and feathers', **3**. The series of plugs are hammered home in cycles to apply an even stress until the block splits away. A clean, vertical break of a few metres can be achieved in this way, more than the bed height of most freestones. Small charges of low explosive are occasionally used to assist in freeing a block. Once on the surface, the quarrymen square the block using plugs and feathers and kivels.

2.06 In some quarries a chain saw is employed to make the vertical cuts to improve the yield. Sawing always appears attractive to block quarries and methods vary with wire and chain saws, **4, 5**. Work is also proceeding with diamond circular saws, and experimentation is continuous. Channelling machines are in use in the Midlands to cut joints, but this can be a laborious process, and the system is principally applicable where there are numerous natural joints. Thermal lances have no application in the limestone/sandstone or slate quarry, and will be described later in the notes on granite extraction.

2.07 Water jet cutting, operating at 275 MN/m^2 has been used successfully on slate and some other stones, and research into this system is being intensified in the UK, the US and Japan. There may also be applications for ultrasonics and lasers in the quarry.

2.08 Not all quarrying is on open faces. Obviously when useful building stone is at a considerable depth below the surface the cost of removing the overburden would make open quarrying completely uneconomic. Chilmark, Reigate (Godstone), Beer, Corngrit, Corsham Down, Aldhelm and Monks Park have all been extensively mined. A typical mine consists of an adit (access tunnel) made in the side of a hill at the level of the good stone. Branch tunnels, up to 3 m in height, are formed off this main gallery sufficient to expose the bed height of useful stone.

2.09 Sansom coal cutters have greatly improved the rate of extraction. These coal cutters can adjust the height of cut from 150 mm to 2 m. The stone is hauled to the surface during the months April-October; during the winter block is stored underground, where the temperature and humidity are constant throughout the year.

Splitting stones

2.10 Some sedimentary stones which can be split freely and readily into thin, easily handled pieces are very suitable for roofing slabs. Limestone of this type will usually occur in beds $\frac{1}{2}$-2 m thick and may be at a considerable depth below the surface. The famous Stonesfield slates were mined on the fringe of the Cotswold area some four miles west of Woodstock at about 20 m below ground. Today, however, the only limestone 'slates' quarried are from north-east Northamptonshire, at Collyweston. The traditional practice is to quarry the stone in the autumn and then to expose the quarried 'logs' to the frost. The slabs, or logs, are dug out and laid over the ground and kept watered. A hard frost followed, ideally, by a sudden thaw is sufficient to allow a skilled man to easily cleave the stone along the bedding and trim it with a hammer. Recent mild winters have caused considerable problems, and thought is being given to the practicalities and economics of splitting the logs by wetting and freezing artificially.

Seasoning

2.11 Stone fresh from the quarry or mine contains a certain amount of water which, on drying out and evaporating at the surface of the block, may deposit some of the natural cementitious matter in the surface pores. This may result in a surface hardening. Drying out also renders the stone more difficult to work. Hardened skins formed in this way cannot really be described as protective. Especially in the case of limestones and calcareous sandstones, such skins may quickly be removed by rainfall. The chief benefits of seasoning are that any potentially damaging salts in solution will, after deposition at the surface, be removed by dressing, and that seasoned stone will not be so susceptible to frost damage.

3 Granite

Available block
3.01 A good dimensional block granite quarry should produce blocks of uniform colour, free of imperfections and not fragmented too small by natural fracturing. Large blocks of 8-10 tonnes should ideally be readily available without the quarryman having to remove too much waste. An ideal quarry of this kind, with reasonable access, is not readily found and the granite trade has to consider constantly the commercial viability of every potential quarry project.
3.02 There are only a few quarries still extracting dimensional blocks in the UK, as distinct from sites where stone is blasted for ballast, and apart from the excellent pinky brown granite from Shap in Westmorland these are all to be found in Cornwall, producing various shades of grey.
3.03 Granite masses may be deep and progress has to be made downwards towards the most useful rock, sometimes to very considerable depths. Where weathering and perhaps glacial action has exposed useful masses lateral as well as vertical development may take place. Sometimes the mass occurs as a hill, and in this case a long open quarry face may be worked.
3.04 Jointing, and the direction in which the block may readily be split are fundamental to the ease with which extraction may take place, to the size of block which will be available, and to the amount of waste involved.
3.05 Generally the finest homogeneous granite masses are found deep down, while masses nearer the surface are cracked and fissured. Extracting granite from a surface quarry necessitates establishing the jointing system and easing the selected piece away from the main face. The size of such a piece may vary between 20 and 200 tonnes. The experienced quarryman confronted with a 200 tonne piece uses a low explosive to lift away a piece (up to 3 m thick) from the face without shattering, leaving it ready for further splitting. The quarryman will, however, frequently have to be content with pieces of 10-20 tonnes of irregular shape which must subsequently be dressed off reasonably square to be of commercial value. Such a procedure may well be costly in terms of waste proportion to useful stone. Much surface quarrying produces not only a large amount of waste, which has to be removed and dumped or crushed, but involves some uncertainty about the percentage of usefully sized block.

Extraction
3.06 Drilling, blasting, and splitting and wedging are all traditional methods of extracting block. More recently the thermal lance has been used to cut a 100 mm wide channel vertically and horizontally and then to free the end of the block, **6**. It may also be used for lance channelling, though this is expensive. There is, however, little wastage, and much saving in time and labour.
3.07 As with the quarrying of other stones, the architect considering the use of granite for a project should ensure well in advance that the quarry he is interested in is capable of producing enough good granite within the required period, as stocks held at quarries are quite small.

6 *Thermal lance with energy equivalent of about 400 kW.*

4 Slate

Sequence of extraction
4.01 The layout of the quarries is dictated by the run of the slate beds. This, with the associated cleavage and 'pillaring planes', governs modern working. Where the beds are on the side of the mountain and lie vertically they are worked back, then succeeding levels below worked back in turn, leaving space at the foot of each face for access and transport, so forming a series of terraces or giant steps. A number of such faces can thus be worked at one time.
4.02 Vertical beds which crop out in valleys can be worked by a progressive sinking of pits with appropriate sides worked into the same terrace form. A combination of the two occurs where depth of working is increased beyond the original floor of the first type of quarry.
4.03 Another method involves mining the slate where the beds descend at an angle into the mountain, increasing the overburden to an extent that prohibits removal. Inclined adits follow the beds down, and at intervals horizontal tunnels are driven in the beds sideways at successive depths. From these tunnels large workings can be opened at intervals with solid supporting walls of rock left between.
4.04 In France to reach slate at depth a method has long been used which involves sinking a vertical shaft to the bottom of the required slate and working large chambers upwards by extracting the slate from the roof.

Extraction process
4.05 The actual extraction of the workable slate in each case was for many years carried out by rockmen, usually in pairs, operating on the rock faces and working off blocks of slate as large as was convenient and allowed by the structure of the material, using natural joints and lightly charged shot-holes on the cleavage and pillaring planes. The cleavage, a distinctive feature of slate, is the plane which splits relatively freely and produces the 'face'; pillaring is a line of splitting lengthwise at right angles to cleavage. Natural joints in the third direction are most useful if not too frequent.
4.06 The art, developed by experience, was to loosen the rock with the minimum of shattering. Gunpowder is used for its softer action in these cases. Gelignite is used for development work and for clearing hard rock. In mining

there is some use of coal-cutter types of machine to give free sides from which to work out the rock.

4.07 Much primary extraction is now achieved by large-scale blasting. Carefully calculated series of deep drill holes will bring down whole sections of rock face, many hundreds or thousands of tonnes at a time. The nature of the rock has still to be borne in mind, but even with a higher wastage it is more economic through speedier production and savings in skilled labour.

4.08 Notably in the Lake District, extensive undercutting with the wire saw, followed by blasting in chambers made in the rock can bring down vast quantities of rock. The wire saw is a very long endless wire, guided by pulleys and cutting into the solid rock with the aid of sand and water—slow but effective.

Shaping

4.09 Once the rock is down on the quarry floor the blocks are trimmed and reduced in size for transport to the works where they are to be prepared for their eventual purposes. Much of the breaking down of blocks can often be done with the plug and feathers (a wedge and split rod) inserted into a carefully placed drill hole on the pillaring and hammered down until the block splits from end to end.

5 Specification

5.01 Specification should be done early enough for modifications which the supplier will advise on to be agreed. These will relate to the choice of stone sizes appropriate, to what is available in the quarry, and to making maximum use of machine working rather than less economic manual work so that production costs can be reduced. The cost of removing material can be high, and removing material has to be paid for by the customer. Early consultation will thus avoid wasteful detailing, such as that involved, for instance, by the use of hollow quoins.

5.02 Suppliers will always welcome the chance to quote for work on the basis of development drawings, which means that the most economic price can be given. Of course, a company will be reluctant to embark on detailed and costly development work unless there is some assurance that it is likely to receive an order. If a nomination is not possible, an order for drawing office work only may be acceptable.

Mould cutting

5.03 Detailed drawings, **7**, prepared by the stone company's drawing office are passed, after the architect's approval has been obtained, to the mould cutting section. Here a card is prepared and issued for the manufacture of each individual stone, each stone being identified by a number, **8**. These cards carry a three dimensional sketch of the stone, or, where required, zinc templates are prepared for both the face and bed profiles of the stones which accompany the cards. Cards and templates are sent to a works control office where a fixing programme and sequence are determined, and the individual stone cards assembled into convenient lots before releasing to the works for production.

Block selection

5.04 The first process in the working of the masonry is the selection of appropriate quarry blocks, **9**. Block sizes vary considerably from quarry to quarry. Massive granite slabs may be made available. At Portland, a piece of limestone ashlar 500 mm high by 1 m long would be described as small, and a stone 1 m high by 1·5 m long described as large. Maximum block dimensions of Springwell sandstone are 1 m high by 2 m long, and of Darney sandstone 1·25 m high by 1·5 m long.

5.05 Within limits, the more freedom a masonry company can be given to specify stone sizes, the more economic the

7

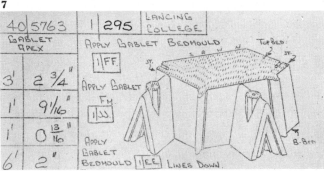
8

9

7 *Preparation of details based on architects' working drawings.*

8 *A card for an individual stone.*

9 *Quarried blocks identified by coding and size.*

job is likely to prove in time and material. In many cases, random course lengths can be very attractive, and enable maximum utilisation of the block with a minimum amount of sawing to produce the required ashlar. There is obviously a temptation to use a fixed module for stone sizes and while this is attractive from a production point of view, it may involve considerable problems with block usage. A notable exception is the practice of some masonry companies in recent years to offer standard facing blocks in three or four standard course heights, usually with a sawn or split face.

5.06 If a fixed module is necessary, for cladding panels for instance, the masonry works and quarry have two alternatives. Either a large volume of block must be produced from which a suitable size will be selected, or an unsuitable block size will have to be used. These alternatives involve either the financing of very large stocks or the acceptance of high material wastage. The careful selection of block for both quality and appropriate size is one of the most important operations, so it is a great advantage to have the quarries and masonry works nearby.

Processing and site works Technical study 1

Workshop practices

Masonry works range from basic slabbing to sophisticated production lines. JOHN ASHURST* describes the processes along such a line: primary slabbing (to give two parallel sides), secondary sawing, machining into more elaborate masonry and finally hand working.

1 Primary sawing

Wire saw
1.01 Wire saws, available with single or multi wires, will cut stone block mounted on a bogie to desired size. Water and abrasive are fed along the wire which can give a very fine, though expensive, finish.

Circular saw
1.02 Two primary saws are mainly used: the large circular and the frame. A primary circular saw consists of a heavy bridge section and a blade, 2 to 3·5 m in diameter, which traverses this, **1**. The blade, impregnated with diamond segments, cuts about 0·25 m² of limestone per minute; sandstone and granite may be considerably slower.

1.03 Block is mounted on independently operated tables allowing the saw to be worked almost continuously. Control of cutting speed is maintained by a closed feedback loop: the measured load on the saw blade controls the stone feed speed to give a fairly constant blade loading. The saw will normally be used for:
- rough squaring of irregularly shaped blocks;
- providing varying slab thickness from the same block;
- slabbing to course height prior to frame or secondary sawing;
- slabbing to thicknesses greater than 150 mm.

Frame saw
1.04 The reciprocating frame saw is used for primary sawing carrying between 10 and 25 blades, tipped with diamond segments, which are each subjected to very high tension, **2**. The reciprocating blades (in principle like a hand saw) are lowered at controlled speed, about 400 mm per hour for example for Portland stone, giving a high standard of finish. This speed is controlled automatically, as for the circular saw. Too great a speed would deflect the blades and cuts would be inaccurate.

1.05 It is slower than a circular saw, though this is partly compensated for by the sawing length per blade being twice that of a circular saw and because multiple cuts are made. The saw is normally used for:
- a higher standard of surface finish;
- greater accuracy;
- slabbing less than 150 mm.

1.06 Other frame saws suitable for less resistant stones rely on teeth for cutting. Granite yards may use toothed blades in conjunction with silicon carbide grit or fine steel shot. Another reciprocating blade is a corrugated steel blank, without teeth, used with steel shot fed in with water which gouges the stone, leaving a characteristic finely scored surface. This is used on some tougher sandstones.

* Based on information supplied by J. V. Borthwick, G. H. Reynolds, J. M. Reid, A. D. Fordham, T. Germain.

1 Primary sawing of large limestone block; 3·5 m diameter blade.

2 Reciprocating frame saw reducing limestone blocks to slabs.

2 Secondary sawing

2.01 After primary sawing stone is generally in the form of slabs 75-150 mm thick sawn on two sides. Further cutting is planned to avoid unnecessary waste. Two independently adjusted circular blades are used and by rotating the table through 90° all four sides can be sawn easily, **3**. Some tables can be operated to rise and fall to allow 'step cutting' and checks and rebates.

2.02 Recently gantry mounted saws have been introduced which speed up processing and can be automatically programmed to undertake a series of cuts to produce stones of required size, **4**.

Slate

2.03 Blocks delivered to the works to be subdivided are split on the cleavage, **5**, and width obtained by splitting on the pillaring line, or sometimes by sawing. Sawing is generally used for cutting to approximate length.

2.04 Blocks are reduced to comparatively small sizes with at least one sawn end and then cleaved with a thin broad-bladed chisel. The blocks are successively halved till the required thickness is reached, **6**. Roofing slates vary from 4-8 mm thick, the thinner slates being generally the more regular and higher priced.

Sawn stone

2.05 The completed masonry, sawn on six sides, may be ready for paving, or with dowel holes and fixing slots, for simple ashlar (ashlar is masonry work in which stones are accurately squared and dressed to given dimensions to make good joints over the whole of the touching surface). Alternatively it is further machined or hand worked.

3 *Secondary sawing of marble slab cutting two surfaces of a sill with two independently adjustable blades.*
4 *General view of Portland masonry works showing primary, secondary and gantry mounted saws (at left).*
5 *Cleavage splitting of large slate block.*
6 *Halving sawn slate block.*

3 Machining

Universally adjustable saw/grinder
3.01 The machine, fitted with saw blade or grinding wheel, is universally adjustable for angle and direction of cut. The stone is fed at controlled speed on a table up to 2 × 3 m and on simpler machines clean rebates and chamfers are made. More complex multi-headed versions may combine up to six operations in one pass, such as angle cutting, surface grinding, slot and groove cutting, and drilling, **7**.

Planers
3.02 Large planing machines designed principally for accurate surfacing of large metal castings are used for continuous runs of complicated moulding such as strings, mullions and sills. Accurate profiles and thicknesses are produced. Planing is carried out dry on a table 2·5 × 6 m, **8**.

Routing machines
3.03 Large routing machines with universally adjustable heads and tables designed for woodworking are useful for soft stones such as Bath limestone. Curved or circular patterns can be cut easily, also vertical turning of large stones and mortising.

Lathes
3.04 Modified heavy-duty metalworking machines are suitable for columns and other circular work. Large lathes are capable of turning up to 2 m in diameter and 3 m high.

7

8

9

7 *Universally adjustable saw/grinding machine capable of angle cutting, slot cutting, grinding and drilling.*
8 *Planing machine preparing long run of limestone mullion.*
9 *Polishing limestone slab with Jenny Lind, a rotary disc on universal head.*

10

Hydraulic splitting machines
3.05 In recent years hydraulic splitting machines have been introduced into the stone trade for walling stones. Sawn slabs are split rather than secondary-sawn to produce a natural split face.

Polishing machines
3.06 For many years the only available polishing machine was a rotary polishing disc on a universal head, called the Jenny Lind, **9**. Though this is still operated, automatic polishing machines have largely replaced it and promise to limit the high cost of polishing, **10**.

Dressing slate
3.07 On some occasions slate is still 'dressed' by hand by chopping down to the required line against an horizontal iron blade, **11**. Now it is mostly done by machines with revolving knives, the slate being placed against a gauge to give required size. With practice the machine cut gives the familiar chamfered edges of dressed slate.

10 *Automatic polishing of marble slabs with single-head polishing machine.*

11 *Hand chopping and dressing slate.*

11

4 Handworking

4.01 Stone may be passed for handworking either to clean down a rough finish or to provide a textured surface (or very occasionally for hand carving). Although most masonry works are now equipped with mechanised hand tools, processing of this kind remains expensive and is therefore employed on a limited scale.

4.02 Smooth finishes are worked with a chopping hammer or a cold chisel, sometimes mounted on a small pneumatic hammer. Rough surfaces are produced by working over the stone with a pointed pick or punch, **12**. A fine punched effect commonly used on unpolished granite is achieved with a surfacing machine operating a four-point chisel.

4.03 Compressed air and abrasives, usually silicon carbide, are in common use in monumental masonry for texturing and are sometimes employed in a masonry works for special effects and carving. The work is carried out in a glass cabinet to confine the abrasives. Rubber stencils protect areas of stone which are not to be subjected to blasting.

4.04 In all those operations the skill of the individual mason is of great importance. Banker work (use of hand tools on the bench) remains a vital section of the masonry works in spite of increased mechanisation. There is always a demand for repair and replacement to traditional buildings, some of which may demand very considerable skill from the banker mason or specialist stone carver. Most work of this kind can be only partly processed mechanically.

4.05 Partly roughed out stone is marked up with bed moulds, face moulds and sections, **13**. The stone is worked first with a 4-5lb hammer and pitcher, and second with a hammer and punch. Punching down by a skilled mason can be to within 5 mm of the finished surface. Final surfacing is carried out with a mallet and sharp chisels, **14**, or hammer and chisels for granite.

4.06 Less tough stones such as those quarried in the Bath or Cotswold areas were traditionally roughed out with a masonry hand saw, and the finish was produced by drags and combs, **15**. Traditional practices are still carried out in small masonry works around the country.

5 Storage and handling

5.01 Stone from the workshops must be stored and handled carefully. Reference has been made already to the need for drying out before fixing (seasoning, see Quarrying TS2). Stones should be protected from rain and frost and stacked to allow circulation of air.

12 *Pneumatic hand tool used for surface texturing granite facing slab.*

13 *Zinc template used to mark out section of cornice.*

14 *Mallet and chisel used in final surfacing.*

15 *First year apprentice rasping Bath Stone.*

Processing and site works Technical study 2

Site works and cladding

The final finish and permanence of stone depend largely on the skill of masons working 'on the wall'. J. V. BORTHWICK describes the on site work of fixing stone, considers tolerances and indicates new directions in cladding design.

1 Introduction

1.01 In the past masons worked the stones at the site, selecting and fixing as the work proceeded. This practice can still be seen in our larger cathedrals where permanent teams of masons are continually employed on repair and restoration. But nowadays the stone is usually worked at the factory, maybe hundreds of miles from the site, and while this puts a high premium on the skills of the draughtsmen and mouldcutters, it has to some extent reduced the standards required of the site operative, who now becomes an assembler of pre-prepared masonry. His skills now consist of the ability to erect this masonry quickly, accurately and safely.

1.02 This has resulted in the formation of small, well organised teams of masons and labourers, often self employed, although most masonry firms have their own nucleus of directly employed labour. In the United States, the stone setter is recognised as a separate trade and it is unusual for the firm which makes the masonry to fix it. Fixing labour is the biggest single item in the unit cost build up of a completed masonry contract and the industry is constantly seeking ways of reducing the proportion of this rapidly escalating cost element.

Site operations

1.03 Site operations consist of unloading, hoisting, bedding, pointing and cleaning down. The stones are delivered to site loaded in boxes or trays or loose packed in straw. Bedding is usually with manually operated chain block tackles, though lighter stones may be lifted by rope tackle.

1.04 Traditionally, stone faced walls with brick or block backing have the stone laid first, backed up course by course with the inner skin. For load bearing stone skins, joints should be raked out as work proceeds and pointed as the job is cleaned down. Where the backing is concrete, fixing proceeds well behind pouring.

1.05 The fixer (walling mason in Scotland) uses spirit level, plumb rule and line, small crowbars and a large wooden mallet to get blocks into position. Strips of metal known as screeding irons are used as spacers to give correct thickness to bed and to horizontal joints.

Staining

1.06 Saturation of masonry during erection is undesirable in every way, not least through causing staining as it dries out through the finished face, depositing alkaline salts from the backing. To prevent saturation stone should be protected by waterproof sheets during erection. Lesser staining and marking could result from site operations so a coat of slurry is maintained on the face until the final clean down.

2 Fixing Methods

Traditional fixing

2.01 As recently as 40 years ago the cladding of a building frame was still largely an extension of traditional masonry, and included bonded ashlar or brickwork and often the fluted columns, enriched capitals and entablatures and pediments of the classical orders involving massive stones, **1**. Even today examples can be seen, as in the Britannia Hotel in Grosvenor Square. Since 1945 stone cladding has tended to become thinner and thinner until the average thickness is now in the range of 75-100 mm, and this has necessitated improved methods of fixing to support the dead load of the cladding itself as well as the applied loads imposed by wind and structural movement (which in earlier days were supported by bonding and the sheer massiveness of the masonry). This has resulted in an increasing reliance on metal fixings such as corbels, cramps and dowels.

2.02 However, metal fixings are expensive and form a significant percentage of the unit cost of masonry. A study of this by the architect, engineer and stonemason can produce substantial economies (see Section 6, Specification).

2.03 Of course, metal fixings have been used throughout the ages. In the Great Temple at Baalbeck, iron dowels 12in in diameter were used to locate and strengthen the column shafts. In the Colosseum in Rome, copper cramps set in lead were used, and mediaeval stonemasons made frequent use of iron ties and cramps especially during the perpendicular period. During the renaissance iron cramps

1 *Sandstone cladding and enrichments of structural concrete frame giving the appearance of traditional masonry. Entablature supported by projecting rc nib.*

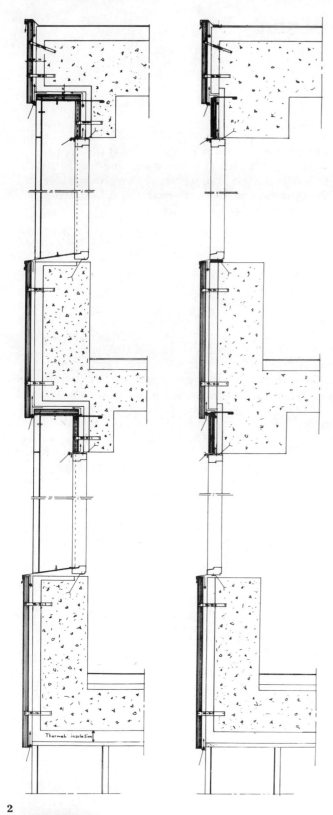

and ties were used, often to support heavy entablatures and pediments. Corrosion was an ever present problem, often leading to bursting of the stone, and cramps were set deep in the masonry to prevent this. Sir Christopher Wren recommended that 'no iron should lie within 9in of the air'. Slate dowels were often used and even oil impregnated oak.

Modern fixings
2.04 Modern thin claddings make extensive use of metal fixings which are invariably made of non-ferrous materials such as copper, phosphor bronze and stainless steel. These can be divided into load bearing and restraint fixings.

Load bearing fixings
2.05 Support of the dead load of the cladding can be achieved by:
- the use of bonder courses
- nibs in the structural frame
- metal corbels (either plates or angles)
- suspended brackets to support soffit cladding.

Restraint fixings
2.06 These locate the stones relative to each other and resist direct or suction wind loads. They generally consist of cramps and dowels secured back into the structure either by means of slotted inserts or by fishtail lugs into blockwork or brickwork. Recommended alloys of stainless steel, phosphor bronze and silicon aluminium bronze for both load bearing and restraint fixings, **2, 3**, are given in CP 298[1].

Costs
2.07 It is difficult to generalise on the economics of the various methods. For instance, the use of projecting structural nibs to support cladding may appear to save expensive metal corbels, but the saving may be counterbalanced by the cost of rebating the stone and casting the nib. Every scheme must be considered on its merits to produce the most economical solution which provides adequate safety factors and complies with the building codes and regulations. Amongst other things these stipulate a minimum thickness of stone behind a cramp mortice which, in the case of limestone, is half the overall thickness with a minimum of 37 mm[1].

Tolerances
2.08 Deformation of the structure under load, shrinkage, thermal and moisture movements can all impose substantial stresses on any cladding units which are rigidly fixed, and the design of fixings should ideally allow freedom of vertical and lateral movement between the structure and the unit. Dimensional changes in steel framed structures may result from elastic deformation under load or thermal movement, but in practice it has been found that these changes are not very different from those of the cladding and normally special jointing techniques are not necessary.
2.09 In concrete framed buildings, however, additional dimensional changes can result from drying shrinkage and moisture movement of the concrete, and creep of the concrete under sustained load. Of all these the most serious is likely to be drying shrinkage particularly if the concrete is cast in situ, although dimensional changes will result from a combination of all these factors.

2 *Granite cladding on British Embassy, Stockholm. Typical Swedish system with air path between slabs and thermal insulation around concrete framing, and using stainless steel cramps and dowels.*

3 *Stainless steel fixing cramps securing slate cladding to insulated concrete backing.*

2.10 It is essential that the design of the cladding and its fixings, joints and beds allow for the building to shorten without exerting compressive forces on the cladding, otherwise these forces may fracture the horizontal supports or the stones themselves, break the ties and cause the cladding to bow out from the main structure. Ideally, to avoid the effects of drying shrinkage it is desirable to delay fixing the cladding if possible until the structure is complete. In practice this is seldom possible.

2.11 It is therefore customary and in most cases essential in a multi-storey concrete framed building to provide compression beds in the stone cladding at each floor level immediately below the cladding supports, be they corbels, nibs or bonder courses. These beds should never be less than 13 mm, and left completely free of mortar or any projecting fixings. The bed should be left open for as long as possible and finally waterproofed with a strip of compressible material, pointed with a sealant such as polysulphide. Expansion or movement joints should be similarly dealt with. The subject is dealt with more fully in CP 298[1, 2].

2.12 Where slate flooring is laid on a concrete screed with cement/sand mortar, uneven riven rather than sawn material may be used and in this case wide joints should be used to accommodate the variations.

3 New developments

3.01 If stone could be obtained in sufficiently large sizes to allow floor to floor cladding panels to be made, much site labour would be eliminated, and fixing of stone would be comparable in cost with, say, storey height concrete cladding. Unfortunately, the natural stone formations seldom allow such sizes to be produced, and certainly not in British limestones. The American Indiana limestone formations do produce such stones and a number of buildings have been completed with storey high stone panels. Because of the high cost of the traditional methods of fixing stone new processes and methods are constantly being developed.

Thin veneer on precast rc

3.02 The technique of backing thin stone veneers with reinforced concrete was first developed seriously in 1960. Since then it has become accepted practice and a number of important buildings in Great Britain have been clad by this method, which is now widely used in Europe and USA.

3.03 Oolitic limestone is ideally suited to this technique as a chemical bond occurs between the calcium carbonate in the stone and the silicates in the cement, as well as a physical bond into the pores of the limestone. However, for complete security stainless steel pins or clips are used to give a mechanical attachment between stone and concrete. The stone veneer is usually 40 mm thick. The individual stones are laid in the concrete shutter, face down, with joint spacers as required. The stainless steel pins or clips are inserted in pre-drilled holes in the stones, the reinforcement cage put in position and the concrete poured. After curing and demoulding the joints are pointed, or left open as desired. The units are then transported to the site and erected in the same way as concrete cladding.

Veneer on lightweight backing

3.04 Use of precast reinforced concrete faced with thin stone veneers has led naturally to the development of thin stone veneers backed with suitable lightweight materials to produce storey-height panels combining the appearance and durability of natural stone with minimum weight, **4**. A very promising material for this purpose is glass reinforced cement, using alkali resistant glass fibre, and development work on this application is in an advanced stage. It is expected that this technique will enable

4

5

4 *Thin York stone cladding cast into factory produced storey-height concrete cladding units.*

5 *Stone panel assembly tensioned by wires. This technique greatly increases the tensile strength of the assembly.*

complicated profiles, which would be extremely expensive if made by traditional methods, to be produced more economically, and may have some interesting applications in restoration work.

3.05 On one major restoration project, serious thought has been given to the replacement of a massive cornice by planing and cutting the moulded profiles and backing these with grc to form a hollow section many times lighter than the original, thus obviating the very serious engineering problems of replacing the feature in solid stone.

Assembly by tensioned rods or wires

3.06 Pre-assembly of stone by the use of tensioning wires or rods is another very promising technique, used already on large projects in London. The backs of the stones are grooved and the stones laid together with resin joints, and wires placed in the grooves, tensioned and then encapsulated in a resin grout, **5**. Alternatively stainless steel reinforcing rods can be laid in the grooves and grouted. When the grout is cured the assembled panels can be transported to the site and erected using techniques similar to those for large concrete units. A development of this technique will clearly be in the direction of increasing the strength of stone in tension, and may increase its scope as a structural material.

References

1 British Standard CP 298:1972, Natural stone cladding (non-loadbearing) BSI, Gr7.
2 Recent experience has shown that dimensional changes caused by sustained load over a period of years can be even more serious than drying shrinkage.

Training for stone Technical Study 1

Training; organisation and apprenticeship

Training of stonemasons provides the competence on which architects' specifications rely, and training of architects should enable them to specify stone realistically and economically. D. MAXTEAD JONES, Secretary of the Standing Joint Committee on Natural Stones, describes the background, and AUSTIN SILCOX CROWE, Chairman of the SJCNS Training Sub-Committee and Head of Training Services for Local Authorities in London, writes on the state of training for the industry today. The authors wish to thank Ted Bedford for the information he provided.

1 Background: organisation of the industry

1.01 The natural variety of stone has had much to do with the somewhat disjointed organisation of the stone industry; this, combined with competition from cheaper and lighter materials and changes in fashion, has played a part in the way stone has fallen behind the market in substitute materials.

1.02 The past few years, however, have seen the ranks of different organisations representing stone closing together. A large number of contractors have recognised the dangers of a divided industry and those represented through their Associations or Federation on the Standing Joint Committee on Natural Stones have been involved with architects and others in examining every aspect of using stone for building. One of the Committee's activities, through its Training Sub-Committee, has been the preparation of a comprehensive training scheme for discussion with the Government's Training Service Agency; another has been to organise short courses for architects to supplement their minimal training in the use of stone.

2 Masonry training

2.01 There is evidence of a resurgent interest in the use of natural stone, but the industry has been losing manpower for several years and is now hardly replacing normal wastage. The loss of manpower is due to a number of factors, all of which affect training facilities, training opportunities and training effort. These factors include:
  nature of the activity
  reduced use of stone for building purposes
  lack of succession planning
  high, and increasing costs of employing and training young people
  lack of continuity in work orders.

2.02 To cut back on training is always tempting because of the costs involved, but in the long-term such a policy is against the interest of the workforce and frequently results in inadequate training facilities at a time when they are most required. The costs of present day training make the situation even more problematical because, despite the fact that the industry's work patterns and structures have changed, there has been no corresponding change in the type and style of training. The modern diversity of activities, although carried out on a much smaller scale, and frequently in a different manner, are still said to demand the same duration and depth of training as before. There appears to be little justification in 'overtraining' to this extent and this becomes especially noticeable when the organisation and sequence of present day work activity is examined. The net effect of this situation is that many employers have ceased to train or reduced their commitment because of the expense involved. If the industry were to review its requirement for training by gearing it to trade/skill activity, commitment to training effort would increase because the system would be more closely tailored to organisational requirements and sufficiently flexible in variety, depth and duration. At present basically one training system exists namely the City and Guilds of London Institute Craft and Advanced Craft, which is expected to be 'all things to all men'. The continuous decline in the numbers of skilled men required has caused many Colleges to close down their masonry activities. In consequence the geographical spread of those centres still operating is such that the indirect costs to some employers have increased enormously because of the need to place their employees on residential courses. These two factors—inappropriate training opportunities and increasing costs—are in danger of combining to further reduce the training effort of the industry.

2.03 The structure of the industry does little to encourage promotion of training as at present provided. The greatest hindrance in this respect is the size of firm—thus not only is there a need to examine the depth and variety of training available, there is also need to organise and present it in such a way that it caters for the needs of the preponderance of small firms. Attempts have been made in recent years to foster specialised training and to train older entrants and semi-skilled operatives, but nothing has been done to develop these ideas comprehensively in such a way that employers have a choice of training to suit their needs and business activities. The Training Services Agency and the Industrial Training Boards are in business to assist and to encourage the development of a national strategy. Essentially their role is to encourage industry to identify the requirement for training and indicate how it might be most effectively organised, but it is not their role to promote training. It is, therefore, not surprising to find that the national training strategy is structurally questionable, because according to size, activity or ownership some firms are within the scope of the Construction Industry Training Board or the Ceramics, Glass and Minerals Industry Training Board. All others, including the Public Sector, and Cathedral Organisations come within the purview of the Training Services Agency (non ITB Division). Each has its own policy, and, in the case of the ITB, systems for grant payment for training done. Thus some firms must pay for all their training, whilst others receive financial encouragement for the same training.

2.04 At present, therefore, there is no structured system which links and co-ordinates the work of these bodies with

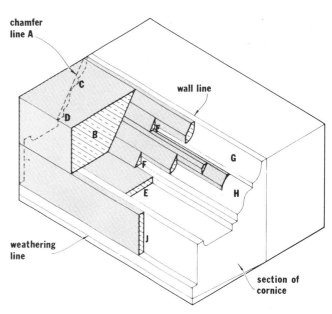

1 Handbook frontis carved in sandstone.
2 Stonemason's tools. Back (left to right) sinking square, mallet, hammer, rasps on top of soft stone drag (used for finishing stones such as Bath). Front (left to right) shift stock, boasters (50 mm chisels for working block to true plane), hard stone chisels, pitcher (for spalling or bursting superfluous stone), punch (for removal of rough, hard stone), scriber and soft stone chisels, driver, dummy and hand saw.
3 First stages in working a true surface.
4 Mallet and chisel used to work stopped moulding.
5 Using a 'second' drag on soft limestone.
6 Finishing a piece of Clipsham stone string course with a rasp.
7 Zinc templet applied to a piece of pediment cornice.
8 Sequence of cutting a cornice. Mark chamfer line A, and work out check B. Produce fillet lines C and D. Work small checks as at E and F. Mark on arris lines for Cavetto mould (about quarter circle) G and hollow of cyma-recta (double curvature) H, and work these curves. Complete mouldings by working small chamfers. Lastly work throat J. The block is cut to the weathering line to produce a fall on the top surface when placed.

specific emphasis on the training of stonemasons. In this respect the industry lacks a national strategy, nor is there likely to be one until:
a) the industry specifies the requirements
b) a more comprehensive system is devised which matches the characteristics of the organisation and structure of the industry, and which anticipates the need to reduce the duration of off-the-job training and caters for the work activities as they are today compared with 50 years ago.

In effect, the industry has to question the premise that all its new employees must undergo a prolonged craftsman course, which is expensive and time-consuming. The answer may lie in fewer highly skilled and expensively trained craftsmen and more people trained (to craftsman standard) in a narrower range of activities as is reflected by the structure of the trades and industry today.

2.05 The need for more training is realised; commitment, due to reasons already mentioned, is lacking. This does not indicate optimism in a period when the trends in increased use of stone are becoming more evident. The prime need in the next three to five years is to anticipate this trend. Failure to do so will result in the architectural profession continuing to find alternatives to stone.

2.06 The need for a national stategy is irrefutable and the present state of training underlines the need for a method of co-ordinating training effort and activity to cover the provision of training facility and to forecast real training requirements.

2.07 A country which takes pride in the maintenance and preservation of its heritage must depend upon having a labour resource which can provide the right balance of skills, including advanced craft skills. These are in danger of being lost because of the ageing character of the workforce and loss of trained skilled manpower. The situation is further compounded by the reduction in the number of training centres. It must be recognised that closure of a training facility results in increased cost to the committed employers and reduction of opportunity to get the type of training required. To this extent it is important to co-ordinate the activities of present centres so that each can make a contribution to the industry on certain specialised forms of training as well as providing the normal 'bread and butter' skills and knowledge common to most of the industry.

Professional training
2.08 There is little evidence of formal training for professionals. Stone has been 'out of fashion' because not enough professionals, especially architects and surveyors, are familiar with its uses and applications, particularly in terms of its cost compared to alternative materials. Post graduate/post diploma training in the use of stone is almost totally orientated to preservation and restoration, and the annual SJCNS short course held at the Institute of Geological Sciences and occasional weekend courses organised for architects and other professionals by the Orton Trust provide the only formal training outside this specialised field.

2.09 To conclude, the industry has the option of continuing to train in a fragmented and haphazard fashion, with the result that committed organisations become more disillusioned because of increasing costs and lack of continuity of work, or it must clearly specify its training requirements in terms of work activity and promote training in a comprehensive manner. To achieve this it should encourage co-ordination of training effort to cater for the needs of craftsmen, semi-skilled and ancillary operatives. Training for masonry activities is in the hands of the industry itself, but in some respects the commitment is dependent upon continuity of work activity and increased awareness of the architectural profession to the need to 'warn' the industry of the opportunities for resurgence.

3 Training stonemasons

Constructional masonry
3.01 Masonry work is divided into two main groups: constructional and monumental. Training described here is for construction. Constructional masonry itself is subdivided by material: limestone, sandstone, marble and granite. Stonemasons often work both limestones and sandstones, but not normally the two tougher stones (which are worked by the marble mason), because techniques and equipment are so different for the two groups of material. Stonemasonry is very much a handtool craft, whereas marble and granite masonry rely far more on machines.

State of stonemasonry
3.02 Though the demand for cladding and speed of erection require organised production and the use of machinery, old methods of production are still employed in various parts of the country. This is chiefly because the local demand for stone has not justified installing machinery. So training, though geared to current trends, produces stonemasons for a considerably traditional industry.

Building Crafts training school
3.03 The illustrations used here are from the Building Crafts Training School. The training programme described here makes reference to the School's outline of the City and Guilds training for stonemasons (shown in italics).

3.04 *Materials*
1 *Recognise and name various stones and state where they are quarried.*
2 *Select stones for given jobs and explain selection.*
3 *Demonstrate knowledge of the care and protection of stonework.*

The stonemason will be aware of the quarrying processes, and recognise faults and potential for fissuring. In particular, bedding planes are difficult to discern in limestone, though easier for sandstone. The ends of natural bedding planes

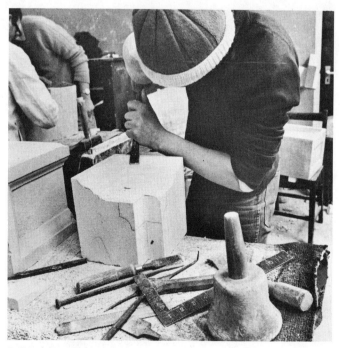

9 Cutting cyma-recta moulding with mallet and chisel.

should be exposed in the building. Cornices and overhanging courses should generally be placed with bed parallel to joint to prevent undercut members falling off during weathering. This sort of intimate knowledge of the properties of stone has become more important because so many specifiers know little about selecting stones in terms of durability, appearance, workability and cost. Many architects rely on the integrity of the mason in the choice and selection of blocks for the various positions in the building.

3.05 *Safety*
1 *Demonstrate knowledge of site and workshop procedures for safe practice when using hand and power tools and machines.*
2 *Know relevant Regulations and British Standards.*

3.06 *Masonry methods*
1 *Work natural stone to predetermined size and shape.*
2 *Measure, prepare and wet mix materials for use in bedding, pointing and grouting.*
3 *Fix natural and artificial stonework from given details and drawings.*
4 *Recognise and name common faults and defects arising from bad workmanship and choice of unsuitable materials.*
Working on block still makes use of simple, traditional tools, **4**. Sawn block is worked to a true surface, **5**, the marked out stone cut to shape with a mallet and chisels, **6**. A fairly fine finish can be achieved on soft stones using drags, **7**; other finishing is done by rasp, **8**.

3.07 *Setting-out*
1 *Interpret scale drawings to obtain the dimensions of given jobs and information required for working drawings and setting-out.*
2 *Set out given jobs to full size and prepare moulds, sections and templets.*
3 *Know practical geometry used in setting-out by full size application and scale drawing.*
4 *Measure existing work and prepare full size and scale drawings for restoration and reinstatement of stonework.*
Once a templet has been used to mark out a stone, **9**, a careful sequence of cutting is followed. The example of a cornice shows the stone to be removed first by working off surplus stone in a series of chamfers and then cutting out a series of fillets to the desired shape, **8**, **9**.

Setting-out in the drawing office
3.08 A stage beyond a stonemason's competence in setting-out is for him to become a setter-out in the drawing office. The setter-out is the link between mason and architect who interprets the architect's instructions and drawings, producing information for the mason. This entails providing working drawings detailing all stonework, verifying site dimensions and getting the architect's approval for all mouldings and other details. From these full-size zinc templets are made. Each stone is given a number which is marked both on the top bed of the finished stone and on the fixing drawing. Often details have to be taken from existing buildings so that old and decayed stonework can be reinstated.

Maintenance Technical study 1

Cleaning and surface treatments

With the increase in the cleaning and treating of stonework with preservatives come the very real dangers of causing irreparable damage to stonework. JOHN ASHURST describes the causes of decay by chemical, physical and biological agents. He evaluates current methods of cleaning and surface treatments and indicates likely harmful side effects.

Causes of decay

1 Water as pollutant

1.01 While 'pure' water is in itself virtually harmless to stone it is nevertheless the greatest enemy of natural stones; it is the vehicle in which the main agents of decay are carried and it is probably true to say that if stone could be kept perfectly dry, most of the causes of decay would be eliminated. Unfortunately there is no known method of completely waterproofing stone without drastically altering its appearance, so that weathering or gradual decay is inevitable in stone exposed to the elements.

1.02 The principal agents of decay can be grouped as:
chemical	crystallisation of salts
	acid attacks
physical	frost
	thermal stresses
	attrition by wind-borne solids
organic growths	algae
	fungi
	lichens
	mosses

Damage and decay may also result from selection of poor stone **1, 2**, or of inappropriate stone for a particular situation, **3, 4**. Bad detailing and simply lack of maintenance may also contribute.

1

3

2

4

1 *Weathering out of a soft clayey bed in a carved limestone pilaster base.*
2 *A fine vent across natural bedding planes has opened up with weathering and been filled with mortar. On a projecting feature part of the stone would be lost.*
3 *Decay of sandstone ashlar below a parapet which was constructed of too porous stone.*
4 *The calcium sulphate solution from the limestone sill is crystallising in the sandstone and causing rapid decay.*

2 Chemical causes

Atmospheric pollutants

2.01 The dramatic improvements caused by controls against atmospheric pollutants have been successful as far as the diminution of sooty material is concerned. But the change in the total emission of pollutants has been minimal over the last 25 years. Stonework subject to atmospheric pollutants will look less dirty than it once might have, but decay will continue at near the present high rate for years to come.

2.02 Carbon dioxide, ever present in the air, dissolved in rainwater will in time dissolve the calcium carbonate cementitious material of limestones and calcareous sandstones so that cohesion is lost and the surface disintegrates, **5**. Sulphur dioxide and sulphur trioxide combine with water to form sulphurous and sulphuric acids respectively which are particularly severe in their attack on calcium carbonate stone. Where Portland stone, for example, is freely washed by these rainwater solutions, the exposed face will generally remain quite white, but it will be extremely rough with the harder fossil fragments standing out in relief, **6**. There will be a noticable absence of biological growth which, apart from some forms of algae, has difficulty in surviving these conditions.

2.03 Where limestone is not freely washed by rain but is only intermittently wetted, **7**, a hard impermeable skin of calcium sulphate will be formed which eventually develops a surface crazing and may lead to blistering or exfoliation of the surface, **8**. Some sandstones are particularly vulnerable to contour scaling. A surface crust, its pores blocked with calcium sulphate for the depth of the 'wetting zone', fractures away from the face parallel with the profile of the face, **9**.

Soluble salts

2.04 Stone may contain deleterious quantities of soluble salts before quarrying and in limestone work may be contaminated by seawater during transportation. Unfortunately the presence of soluble salts can seldom be detected by eye and a perfectly reliable masonry contractor may supply salt-contaminated stone in ignorance of its condition. Some forms of salt contamination, such as those by sulphur acids in the atmosphere and chlorides in sea spray, are unavoidable, **10**. Other forms may be controlled. For example salt-contaminated sand is too often used for mortar and concrete and as moisture evaporates from or near the surface the salts are readily conveyed from the sand to the masonry, **11**.

2.05 All soils contain soluble salts and provided there is sufficient moisture their passage by capillarity into a porous stone is inevitable. Fertile soils contain a great variety of salts, and manures—both natural and artificial—are a prolific source of supply. Salts from the ground have been known to travel to the first floor of a building and in this connection lightning and atmospheric electricity may play a considerable role, **12**. It is known that salts are ionised by the currents thus increasing their activity.

2.06 Where water movement governs the distribution of soluble salts the capillary properties of the stone and the

5 Acidic rainfall may cause rapid deterioration of sandstones whose cementing matrix is principally calcium carbonate. This arch shows almost complete loss of profile in a highly polluted environment.

6 Rough texture of weathered Portland stone. All limestones are soluble in water especially if it contains CO_2. Weathering is often slow and not very serious.

7 The washed skin of these clunch columns remains intact while the intermittently wetted 'protected' side has completely shed its sulphate skin.

8

9

10

11

12

8 Decay may be particularly severe on magnesian limestones. This recently washed tracery shows how the sulphate skin has crazed, blistered and spalled. Cavernous decay often results.
9 Contour scaling of sandstone as a result of contamination with calcium sulphate. Large scales are lost in this way.
10 The result of wind-borne salt spray saturating stone six miles inland. High winds causing rapid drying result in frequent crystallisation cycles and accelerate decay.

11 Damaging salts may be present as the result of using contaminated sand in the jointing mortar or backing. Decay is first seen around the joints.
12 Rising damp evaporating beneath the windows is depositing salts which stain and cause disruption of the surface.

rate of evaporation are critical factors in the ensuing rate of decay. In high ambient temperatures, or on faces of stonework directly exposed to the sun, evaporation is very rapid. Vaporisation starts to take place beneath the surface of the stone and the salts are deposited internally.

2.07 Crystallisation of the majority of soluble salts will set up sufficient forces to cause serious damage to most porous stones. The damage may take the form of exfoliation, spalling or dramatic shattering. In lower temperatures, evaporation is from the surface and the salts are carried through the stone in solution and are deposited on the face of the stone in the form of efflorescence. In such cases their decaying effects are usually far less dramatic but they may corrode the face of the stone if allowed to remain in contact with it. Efflorescent salts can usually be removed without great difficulty but unless the source of the salts is denied, and the ingress of water prevented, the periodic removal of salts from the surface is of little avail.

3 Physical causes

Frost
3.01 Damage from frost is rare in the UK and only occurs in conditions of severe exposure. Limestones are more susceptible to damage than other stones, and this susceptibility is related to pore structure, which governs the degree of saturation and the size of stresses generated during freezing, **13**.

Thermal stresses
3.02 When stone is exposed to hot sun during the day, night-time radiation will reverse the source of heat so that a continual cycle of differential stresses between the surface and the core is set up. In extreme climates this stress reversal may cause fatigue and ultimately fissuring. Although not unknown, it is very rare to find damage of this kind in the UK.

Attrition by wind-borne solids
3.03 Abrasion of decaying stone surfaces may occur in marine zones where high winds carry sand inland. But the type of decay popularly referred to as 'wind erosion' is usually the result of rapid wetting and drying of wind-borne salt solutions. Repeated crystallisation cycles in these conditions may cause very spectacular decay, **10**, which, occasionally is aggravated by wind/sand scour.

13 Splitting off of surface layers through frost action on stones in wet conditions.
14, Attractive as such growth may be, it is often the first step toward larger and positively destructive growth taking place, 15.

4 Organic growth

Algae
4.01 Algae may be green, red or brown powders or filaments which may be slimy according to moisture conditions. They occur on all types of building surfaces outdoors and they will thrive even in industrially polluted atmospheres. Though not usually destructive to stonework they are unsightly and on steps or paving they constitute a hazard to pedestrians. In common with all forms of vegetation on stonework they tend to encourage water retention within the material and for this reason they should, wherever possible, be removed, **14, 15**. Symbiotically, with some types of fungi they lead to the growth of lichens.

Fungi (moulds or mildew)
4.02 Moulds appear as spots or patches which may spread to form a grey-green, black or brown furry layer on the surface. They rarely appear by themselves on stone as they depend on organic material for sustenance.

Lichens

4.03 Of all the biological growths on stonework, lichens are the most difficult to eradicate without the use of substances which might be detrimental to the stone. There are many varieties but those which cause the greatest disfiguration are the calcicolous which, as the name implies, grow on calcareous materials. Light-coloured limestones can become most unsightly as a result of lichenous growths which develop as green, black or pink circular patches. The colouring matter consists of the fruiting bodies and thallus of the organism, and it is the thallus—the majority of which grows beneath the surface in minute irregularities in the stone—which propagates the growth of the fruiting bodies. They are extremely slow in growth and the individual rings usually progress at the rate of 0·5 mm per annum (1 mm in diameter). They can tolerate extremes of temperature and survive for lengthy periods without moisture.

4.04 Lichens have to struggle for survival in urban areas where sulphur, oxides of nitrogen or excessive carbon dioxide are present in the atmosphere. Although the organisms secrete some acid there is no direct evidence that they contribute seriously to the decay of stones. But lichens do change the surface texture and apart from unsightliness should not normally be tolerated on finely carved or otherwise valuable masonry.

Mosses

4.05 The growth of mosses on stone, or in the joints, is usually indicative of abnormally wet conditions and invariably points to the onset of serious decay. Whereas the aesthetic appeal of lichenous growths on rustic stonework in a rural setting might override the marginal danger of water retention and surface disruption, the presence of mosses should never be tolerated on any form of stonework.

Cleaning natural stone

5 The case for cleaning

5.01 Most moderately soiled stone on buildings is cleaned for aesthetic reasons, **16**. Dirt disfigures and obscures the colour and character of different stones, reducing them to an unattractive uniformity. On older buildings a wealth of detail, especially at high level, may be lost completely under black sooty deposits. However, advanced soiling is not simply an aesthetic problem because dirt, as well as hiding open joints and structural faults such as cracking, is also a major source of decay.

5.02 The cleaning of older buildings has been criticised on the grounds that the 'appearance of age' is being removed. But examination of such a building after cleaning will show that the weathering of stone is more than accumulation of dirt. Heavy dirt deposits are an obstacle to natural weathering.

5.03 Cleaning can be carried out in a number of ways: washing, abrasive blasting, chemical cleaning and mechanical cleaning. Choice is governed primarily by the type and condition of the stone and the degree of soiling, and also by the use of the building and its accessibility.

16 *Before* **a** *and* **b** *after cleaning in the inner courtyard of the Royal Academy, London.*

6 Washing

6.01 Traditionally limestone is cleaned by softening accumulated dirt with water sprays. Brushing and scraping assist removal. This is still the simplest and often the best way to maintain the clean appearance of a building. But because so many buildings of every age are now being cleaned for the first time, the quantities of water needed to remove the dirt are likely to have deleterious side effects. So washing must be supplemented by other methods, though it should still always be used on areas of detail. No one would select a cleaning method involving abrasion unless the impracticality of simply softening and washing the dirt away demanded it.

Washing limestone

6.02 Dirt adhering to the porous surface of limestone for a long period becomes attached with a binding matrix of calcium sulphate. Constant wetting of this skin over the years drives dirt into the pores of the stone so that it is almost impossible to wash all dirt out of a heavily soiled surface in one washing operation. The removal of superficial dirt is followed by an almost immediate brown or 'ginger' staining caused by a tarry solution from the pores drying out on the surface. The staining is directly related to the amount of soiling and porosity of the stone and so is particularly a problem on very old buildings. Subsequent washing removes this stain which in any case fades with time. The phenomenon is most noticable on light coloured stones such as Portland and Kent rag, but it is scarcely a problem on brown or yellow stones, such as Ham Hill or Anston. A particular benefit of washing is that it gives 'in depth' cleaning; a dramatic surface clean such as blasting often disappoints building owners because browning follows rapidly. The brown colour is frequently the natural colour of the crystalline sulphate skin which has not been removed by cleaning, **17**.

Water penetration

6.03 A more serious disadvantage is water penetration; at no other time in its life is a building subjected to such a concentration and quantity of water as during washing. Hidden iron cramps and fixings normally out of reach of rain may be saturated and may subsequently rust, stain or spall; cracks and open joints may allow the build up of water around buried timber or behind panelling. Also the relative porosities of stone facing and brick backing may hinder the drying of the backing and lead later to dry rot. A further disadvantage is that washing must take place during frost-free months.

6.04 Most of these objections can be overcome by careful planning, proper preparation of the elevation, constant and conscientious supervision and by using supplementary techniques, such as hand rubbing with abrasive stones, non-ferrous wire brushes, and wooden scrapers, or even a limited amount of blasting on heavy encrustations. Often whole buildings are saturated without need, in the attempt to soften a few stubborn spots, encrustations under mouldings or hard tarry spots in pieces of carving.

Methods of washing

6.05 Soaking with sprays softens dirt deposits and dissolves the soluble binding matter; the softened deposits may then be removed with brushes, scrapers or water jets. Mains water is normally used (even when this is chlorinated). It is applied through fine to medium V-jets, **18**, or cone jets

17a *The Scottish Office in Whitehall;* **b** *cleaned in 1966 by grit blasting; and* **c** *now both stained and weathered.*

18a *Before and* **b** *during washing intricate carved stonework.*

arranged on booms which can be moved up and down the facade as required. These should be spaced to give even saturation of the facade and should be individually controlled to avoid waste on clean areas or unwanted water on windows. Booms may be supplemented by individual spray heads on hoses looped and tied to the scaffold, but this arrangement is not acceptable for cleaning a large facade because positioning is limited and the nozzles tend to slip and change position. In some circumstances an oscillating 'rain-fan' boom may be useful.

6.06 There is no advantage in using coarse rather than fine jets. A fine spray discharging 110 to 140 litres/h is satisfactory but there is a good case for using nebulous sprays discharging approximately 45 litres/h where fine or fragile detail is to be cleaned. Nebulous or 'hydraulic atomised' sprays create a wet mist and substantially reduce the amount of water cascading down a facade. But to be effective they must be clustered close to the dirty surface and tightly screened to prevent the mist spray blowing about.

Steam
6.07 Steam was used quite extensively before the last war but fell into disrepute partly because caustic soda, added to the boiler water to avoid furring, was deposited on the cleaned surface and remained there to cause decay. Because steam condenses so quickly the method is little more than a hot water wash with rapid drying. Hot water is no more effective than cold in removing atmospheric dirt and so there is little point in its use. A useful modern application of the steam lance is the removal of disfiguring trodden-in chewing gum from paving.

High pressure lance
6.08 The cutting action of the high pressure lance, using cold water, is useful in removing stubborn patches of dirt and as an adjunct to both washing and blasting where it is used to freshen up the facade and clear it of dust, **19**, and also in chemical cleaning where it removes both the dirt and cleaning agent. Chemicals, both acid and alkaline, disinfectants, and petroleum emulsions for special uses can be incorporated. Detergents are included in some chemical cleaners where their principal use is as wetting agents. There is nothing to be gained from scrubbing or soaking with soapy water; detergent powders containing sodium sulphate should be excluded.

7 Abrasive blasting

7.01 Early enthusiasm led to some unfortunate mistakes where the use of blasting was very ill-advised; blurring, pitting and loss of surface is still an all-too-frequent sight where blasting has been too severe. However, blasting has proved its worth on heavily soiled masonry, **20**, especially on siliceous stones; it is often the only way to remove the heavy encrustations present on 'first cleans', and so it can supplement washing or chemical cleaning.

7.02 The main attractions of blasting are speed and the immediate and often dramatic overall result. The associated noise and dust nuisance may be thought a small price to pay for these twin benefits, especially as with dry blasting the fears of water penetration and tarry staining are removed. Very good results can be achieved where hardness of stone, boldness of detail and a craft approach by the operative come together. On the debit side, irreparable harm can be done by indiscriminate blasting of soft stone which could have safely been cleaned by water or simply by a change of abrasive and a smaller gun.

7.03 One unfair criticism of blasting is that it removes 'case hardening' on the stone, thereby exposing it to accelerated weathering. 'Case hardening' usually refers to an irregular

19 *High pressure jet used in 'cutting off' dirt softened by other processes.*
20 *Abrasive blasting of heavily soiled masonry. Great care is needed to avoid obliterating the detail.*

crystalline calcium sulphate skin; in time this will craze or erupt in unsightly blisters. Occasionally this skin may well be more resistant to blasting than to water. On two comparable and adjacent areas of Portland ashlar, water saturation dissolved some of the calcium sulphate leaving a rough surface while blasting cleaned the skin, but left it intact.

Methods of blasting
7.04 Blasting cleans by means of a compressed air jet containing an abrasive, either silica sand or non-siliceous grits, eg ground copper or iron slags. Compressed air is fed to a pressure pot containing the abrasive, and the two are then passed along a hose to a blasting gun.

7.05 Various grades of sand and grit are available. Generally speaking very coarse grades are no advantage and tend to block and spurt unevenly from the gun. Finer grades have a smoother flow and usually remove dirt faster unless the stone has a very rough texture. The choice between sand and non-siliceous abrasive is usually determined by cost, for although it is a health hazard, sand is about 50 per cent cheaper than most grits. Relative cutting speeds vary with the type of stone and the operative. There is a variety of pressure pot and gun sizes;

21 *Air abrasive blasting using a 3 mm ceramic nozzle. Great care can be taken with small nozzles and little dust results.*
22 *Effects of acid cleaning on sandstone.*

the smallest types allow the operative to control the spread of abrasive and to use the gun on carved work where arrises would be vulnerable to the wide spray of abrasive from the larger guns, **21**. Fine abrasives only should be used with these small guns because they would be blocked by coarse grades.

Wet blasting
7.06 A mixture of water and abrasive tends to be less harsh than the dry abrasive, but this benefit is offset by the amount of slurry generated at the wall face. This makes wet blasting unpopular with operatives because even when they are properly attired, it tends to obscure the work by covering the plastic window in the helmet and by adhering to the wall face and collecting on ledges and in mouldings. The net result is that a light and dark mottled effect (known as 'gun-shading') and cementitious build ups of dirt and slurry may be left behind.
7.07 An important advantage of the wet blast is that it reduces to a minimum the free dust which can be such a nuisance with a dry blast. A wet-blasted facade should be well washed after completion, preferably with a high pressure water lance, to remove dried films of slurry; build ups of slurry on the scaffold and at ground level should be cleared away each day to avoid blocking gulleys and surface water drains. But even though considerably less water is used during wet blasting than during washing, tarry 'drying out' stains must be expected where there have been heavy dirt deposits on porous blocks.
7.08 Noise may be a more serious problem. Telephone conversations become very difficult in rooms in the blasting area, even with windows shut and sealed, and any effort at concentration is difficult. There is no appreciable difference in noise levels between wet and dry blasting.

8 Chemical cleaning

8.01 Chemical cleaners are usually based on acids or alkalis. Most either contain soluble salts or react with stone to form soluble salts, so the cleaner must be completely removed at the end of cleaning.

Acid cleaners
8.02 The only cleaner known to leave no soluble salts in masonry is hydrofluoric acid though this is extremely dangerous in inexperienced hands. It is highly corrosive so is restricted to removing soiling from sandstones and granites which water cannot touch. It is also used for soft, heavily soiled brickwork which is unresponsive to washing but would be ruined by blasting.
8.03 There are several disadvantages in using this acid apart from the danger to operatives. Glass may be etched and the polish on marble, granite and glazed tiles may be attacked. It may also attack free iron in some sandstones, causing rust staining (a proprietary rust inhibitor will overcome this). The acid is applied by brush or spray to a pre-wetted wall and left for a short time before washing off, being reapplied and thoroughly washed off with a high pressure water jet. This usually achieves a high standard of cleaning and is quiet and relatively cheap, **22**.

Alkaline cleaners
8.04 Alkaline cleaners are based on caustic soda with additives to control penetration and promote surface activity. Their main use is cleaning moderately or lightly soiled limestones and to a lesser extent glazed bricks and faience. They are sometimes used on ordinary porous brick, but this is not advised because more of the cleaner may be absorbed than can be satisfactorily washed off and soluble salts will be left behind. Alkaline cleaners compete with water washing rather than with blasting. Encrustations, heavy soiling, or indeed any soiling which requires more

than two or three applications of the cleaner should be finished by another method. The soiled area should be first wetted, working from bottom to top to minimise the risk of streak staining and each application should be jetted off with clean water before the next is applied.

8.05 The final removal of the chemical by washing is imperative, or efflorescence and bloom will result. The time which any chemical cleaner is left on a building is critical. Hydrofluoric acid preparations left on too long will result frequently in the formation of colloidal silica which shows itself as a white bloom and is almost impossible to remove. There is also an optimum time for alkaline cleaners.

9 Mechanical cleaning

9.01 As with abrasive blasting, the abuses and misuses of mechanical cleaning can readily be understood, but it is still an important method for cleaning sandstones and grit stones. Power tools with carborundum heads, rotary wire brushes, or abrasive blocks (both artificial and natural, such as Derbyshire gritstone) are used on stones soiled with water-insoluble deposits. Mechanical cleaning is best used on flat areas or large-scale simple moulding, and then should be coupled with blasting or acid cleaning on fine detail. Great skill is required to achieve a good finish. Another consideration is cost, for while discing and blasting may compare favourably in price with straightforward blasting, if the surface is to be hand-rubbed the price may be 30 or 40 per cent more.

10 Special cleaning problems

10.01 A guide to cleaning problems is given in table I. Common disfigurements of masonry are rust staining from ferrous window bars and hidden fixings, green staining from copper and bronze, green and black algal slimes, efflorescence, old paint and limewash and oil or grease stains. Soluble limestone dressings can also mar nearby brickwork.[1]

Metal stains

10.02 Long standing metal stains are almost impossible to remove. Some acids may be successful on stains of only a few months' duration and specialist firms will undertake such work. Ammonia solutions are useful for copper stains and for washing off and inhibiting the regrowth of algal slimes. Water repellent liquids and neutralising liquids may be used to combat efflorescence by changing the moisture content on the wall on which it forms, although if practicable, simple brushing off from time to time may be a better method. The use of silicone water repellants is currently being observed (by DOE Directorate of Ancient Monuments and Historic Buildings) on brickwork subject to staining from soluble limestone dressings with the object of improving the self-cleaning properties of the wall and preventing the formation of stains and deposits.

Limewash

10.03 Old sulphated limewash in multiple applications is often too difficult to scrub and wash off; it should either be wet poulticed over a long period to soften the limewash or, if the wall will stand it, blasted with abrasive.

Grease and oil

10.04 Grease and oil stains may be removed using carbon tetrachloride in well ventilated conditions; application is either by sponging or by a series of poultices using whiting or one of the natural clay earth media available.

Clay earths

10.05 The clay earths have the advantage of being easier to remove; there is no pore filling nor any necessity for scrubbing after the poultice is removed. Poultices are widely used, with and without gelling agents and detergents, for cleaning statuary and marble. They could well be used more on carved detail on buildings where prolonged treatment by other methods carries too great a risk.

Organic growth

10.06 Organic growth may become a nuisance, for although only a few low growth forms are directly harmful to masonry, they may be soil forming and create root holds for higher growth. An effective treatment now used extensively by DOE is to double treat with a quaternary ammonium compound, followed by a tin oxide/quaternary ammonium application as an inhibitor. This treatment gives long term protection. The site is *dry* brushed between treatments, and treatment should be on a dry day.

11 Selection of methods

11.01 Even a brief study of cleaning techniques will show how mistaken it is to adhere to any single method as correct especially with the problem of the 'first clean'. Different stones, conditions and amounts of soiling demand a flexible approach to the choice of method, see table II.

11.02 Ideally, all abrasion and saturation would be excluded. Whatever caution at present rightly surrounds the use of chemicals, it is the use of those in the future which may solve most of the present difficulties associated with other methods. It is to be hoped that research in the field of chemical cleaning will eventually overcome the hazards of hydrofluoric acid and the soluble salt problem of alkaline (sodium hydroxide) cleaners. For the time being however, these methods must take their place with the other techniques, to be used correctly or not at all.

Table I Common defects associated with cleaning		
Method	**Defect**	**Cause and notes on avoidance**
Washing Wet blasting Pressure lancing	Tarry stains	Tarry solution, formed during wetting dries out on pores. Can be reduced by avoiding prolonged saturation and by further washing. Some tarry staining unavoidable.
	Dry rot Rust expansion Flooding	The results of penetration can be minimised by careful sealing of all open joints and cracks before cleaning, and by taping and sheeting all openings. Watersheds and catchment sheets on rigid supports with falls to gulleys will avoid flooding risk.
Dry blasting	Pitting of surface	The result of wrong choice of method on stone which is too soft; or a careless operative; or the use of too harsh an abrasive.
Wet blasting	Blurring of arrises	
	Gun shading	Especially on wet blasting erratic movements with blasting gun leaves mottled effect. Slight appearance often unavoidable. Pronounced appearance the result of inexperienced operative.
	Blasted glass	Careless use of gun and inadequate window protection. Glass should be coated with peelable protection.
Wet blasting	Slurry deposits and film	Unfinished job. All dust deposits and slurry should be hosed or jetted off.
Acid	Brown stains on sandstone	Stone with high iron content. Acid combined with rust inhibitor should be used.
	Etched glass	Caused by lack of protection or, if occurring after cleaning, acid vapour on scaffold. Peelable plastic coating should be used on glass and scaffold boards washed and lifted.
	Pavement staining	Splashes of acid not neutralised and washed away.
Alkali	Efflorescence	Excessive number of applications, careless washing off or wrong use on too porous a material.
	Streak staining	First wetting and application carried out from top to bottom. Risk can be reduced by working upwards.
Mechanical	Scour marks	Lack of skill or wrong use of method on moulded stone. Can be improved with hand rubbing.
	Wavy arrises	Lack of skill or more probably wrong choice of method on carved or moulded work.

Table II Methods of cleaning stone—selection guide

Method	Calcareous (eg limestone)		Siliceous (eg sandstone and granite)	
	Degree of soiling		Degree of soiling	
	Heavy	Moderate to light	Heavy	Moderate to light
A Washing*	Yes† B or C on hard stone	Yes	No	Yes† B or C on hard or E on soft stone
B Dry blasting	Yes	No	Yes	Yes
C Wet blasting	Yes	No	Yes	Yes
D Chemical* (alkaline)	No	Yes	No	No
E Chemical† (acid)	No	No	Yes	Yes
F Mechanical	Rarely	No	Yes	Yes

* D additional use on glazed brick, faience etc.
† A and E also used on brickwork.

11.03 There is no justification for taking unnecessary risks on valuable buildings with unproven methods, however plausible the method may seem. An important building which has accumulated dirt for a hundred years or more can well wait to see the method proved on lesser facades. On the other hand there should be no unreasoned prejudice against a method because of one or two bad results, when there is ample evidence of satisfactory cleaning available elsewhere.

12 Recent developments

12.01 Cleaning of natural stone and other building materials is a developing technique. Materials and methods become more sophisticated and efficient and some of the more difficult cleaning problems are now within the range of competent cleaning companies who, a few years ago, might have found it necessary to recommend less than ideal solutions. An early discussion with a cleaning contractor is always advisable where special problems are likely to be encountered, and 'trial areas' are often useful in agreeing a standard before the main work commences.

12.02 A number of chemical cleaners have been developed to remove aerosol paint graffiti, bituminous paints, rust and bronze stains, oil etc, in combination with high pressure low volume water. All the precautions described in the main text apply.

12.03 Recently a medium/high pressure water jet containing abrasive, usually sand, has been developed (distinct from 'wet blasting'). This method is fast, even on fairly heavily soiled sandstone, and is dust free. As with any abrasive system, however, control of pressure and angle of 'cut' is very important, and experience and supervision are essential. A water abrasive vacuum system is also available and may be useful on internal areas where excessive dust and/or water cannot be tolerated, **23**. Further details of some recent developments will be found in TS3.

Chemical treatments
13 Reliability

13.01 Of the attempts to use chemicals to improve the weathering of stone, a few have had limited success, most have proved worthless and some extremely hazardous, (8). There are so many variables—the method of working stone, how it used, climate, concentrations and type of atmospheric pollutants—that any so-called preservative marketed for general use should be regarded with great caution. Important recent developments are described in TS3.

13.02 Since water in stonework is the greatest danger to its well-being, many agents of decay would be eliminated if it could be kept completely dry. Most chemical treatments are designed to keep stone dry and/or to consolidate friable material. A third group of chemicals, toxic washes designed to kill biological growth on stone, are referred to under cleaning techniques. Some treatments combine water repellants and consolidants, others claim to inhibit biological growth as well.

14 Keeping stone dry

Waterproofers

14.01 Earlier comments about the action of soluble salts and frost on stone suggest that there should generally be no half-measures; the stone surface should be completely impermeable. Common substances such as paints and bituminous compounds are ignored here since they destroy the aesthetic appeal of natural stone.

14.02 Other 'waterproofers' and 'preservatives' such as linseed oil, waxes and metallic stearates have been in use for many years. Their intended function is to fill the pores of the stone with a water-insoluble deposit. Linseed oil and wax tends to change the stone's appearance and attract dirt. In certain conditions, some metallic stearates may liberate acid. None of these treatments is long lasting and none can be recommended.

Silicone water repellants

14.03 A small number of suppliers supply silicones to a large number of companies producing silicone water repellant by adding them to organic solvents or aqueous bases.

14.04 BS 3826:1967 (1969)[4] deals with three classes of repellants. Class A is for use on predominantly siliceous stone such as Stancliffe, Blaxter and Hollington. Class B is for predominantly calcareous stones such as Portland, Bath and Clipsham. These classes have an organic solvent which cannot be applied to a wet surface. They penetrate better than aqueous based repellants and can be applied more successfully on surfaces previously treated. The

23 *Water/abrasive vacuum system. The head doubles as air/abrasive nozzle and vacuum hose.*

silicone barrier inhibits penetration of low pressure water but allows water vapour to diffuse through it. Average penetration for medium porosity stone is about 3 mm. The dramatic 'duck's back' effect of water globulation evident immediately after application disappears quickly on exposure and the immediate surface then readily retains moisture.

14.05 Class C materials are for the same conditions as Class B. They are caustic and not recommended for sandstone containing ferruginous (irony) matter which could suffer from rust staining. Class C materials can be applied to damp masonry and so are sometimes used before A or B assuming that the masonry will eventually dry and be conditioned for the absorption of more efficient organic solvent based materials.

Penetration from brush or spray application rarely exceeds 2 mm. When deposited their effect is as for A and B.

Reliability

14.06 BS 3826 and BRS digest 125:1971[5] warn of the danger of damage and recommend expert advice. Stone Preservation Experiments (8) describe detailed observations of these materials on different stones.

15 Consolidants

15.01 Although all consolidants must to some extent inhibit ingress of water this may be incidental. They are designed to desposit a bonding agent in the pore structure of interstices of the stone. Linseed oil, waxes and metallic stearates which provide some consolidation were mentioned in **14.02**. The most popular marketed have been ethyl silicates or siliconesters recommended by reputable suppliers, though not specifically mentioned in BRS digest 125. They are often applied in conjunction with Class A or B silicone water repellants either separately or using the same solvent as vehicle.

15.02 Siliconester combines with moisture to leave a silica desposit in the pore structure. It is effective beyond dispute but its drawback is lack of penetration; seldom beyond 3 mm even after total immersion of the stone.

Repellants and Consolidants with biological inhibitors

15.03 None of these materials are effective 'preservatives' because of shallow penetration. TS3 describes the development of deep impregnants as 'preservatives'.

16 Conclusions

Frost damage

16.01 Silicone and siliconate water repellants are of limited value in excluding water from stone and in temperate and cold climates may actually promote water retention and consequently frost damage.

Salt damage

16.02 Silicone or siliconate water repellants or siliconesters are particularly dangerous to use on materials containing appreciable quantities of soluble salts or in contact with other materials containing these salts[8].

General conclusion

16.03 Is there a place for these chemicals in maintaining stone? They could be used on structures of little historic or intrinsic value to ameliorate the results of water penetration and where the benefit outweighed the risk involved. It is suggested that they should not feature in the conservation of items of great value or historic significance.

16.04 There is no place for chemical impregnation of new stonework; thoughts of doing this are an indictment of the design and the selection of the stone. Improvement of new stone is unlikely unless a method is devised of achieving a uniform distribution of applied materials throughout the stone such as by vacuum impregnation.

16.05 The evaluation of a stonework treatment is extremely lengthy. While accelerated laboratory tests may produce failures quickly, good results are a poor indicator of field conditions because there are so many extraneous factors to be considered.

References

1 Building research *digests* 20 and 21 (first series): *The weathering, preservation and maintenance of natural stone masonry.*
Also of interest, 2 and 3:
2 Building research *digest* 113: *Cleaning external surfaces of buildings* (currently under revision).
3 *Code of practice on stone cleaning and restoration* (August 1974) Federation of Stone Industries.
4 British Standard 3826: *Silicon based water repellants for masonry*, 1967 (1969).
5 BRS *digest* 125: *Colourless treatments for masonry*, Jan 1971, 6p.
6 BRS *digest* 139: *Control of lichens moulds and similar growths*, March 1972.
7 BRS *digest* 177; *Decay and conservation of stone masonry*, May 1975
8 B. L. Clarke and J. Ashurst, *Stone Preservation Experiments*, BRE: 1972.
9 R. J. Schaffer, *The weathering of natural building stones*, 1932 (reprinted 1972), HMSO.
10 AJ Technical Study: *Decay and Cleaning of Masonry*, W. H. Dukes/J. Ashunt, 1972.

Maintenance Technical study 2

Diagnosis and repair methods

Growing public interest in our heritage of old buildings has given architects new responsibilities for repair, restoration and adaptation. REG WOOD describes the process of diagnosis and the methods on which architects can base a repair policy.

1 *Ruins of former archbishop's palace, Southwell. Ruins are vulnerable to the weather and create special problems of support.*

1 Introduction

1.01 Much of our building heritage is of stone: cathedrals, churches, castles and stately houses, cottages, barns and ancient boundary walls too. How does an architect new to this kind of work deal confidently and sympathetically with the repair and restoration of stonework?

1.02 A good stonemason, sadly in short supply, is essential. But this is not enough. The architect must decide a policy for repair according to the building type and condition of the stonework and explain his intentions by specification or by frequent discussion with craftsmen on site. The first step in formulating a repair policy is for the architect to make a methodical examination of the building.

2 Diagnosis

2.01 Repair policy will be influenced first by the type of building and its age. If it is a classical building, the policy will probably be to restore the complete architectural conception. If it is an ancient monument no longer in use, possibly in ruins and preserved only as a 'museum piece', the policy will probably be to retain as much of the original stonework as possible, slowing down if not arresting its decay. When a monument is in ruins special problems of support and consolidation arise and in unroofed monuments special protection is needed to combat weathering, **1**.

2.02 These two general policies, of maintaining a complete servicable building and of preserving the antiquity of the stones themselves, may be combined in old churches and cathedrals.

Inspection
2.03 It is convenient to describe inspection of stonework under three headings: walls generally, joints in stonework and stones themselves.

General condition of walls
2.04 *Are there any cracks, bulges or signs of settlement in walls?* If they need attention, this should precede stonework repairs. Many walls which lean slightly, bulge or have minor cracks may have been in this condition for a

2

considerable time and may not need any repair.
2.05 *Is the wall dirty?* Encrusted soot encourages decay in stonework. Stone cleaning might be necessary for the sake of appearance, for the preservation of the stonework or to ascertain the full extent of repairs needed. (See Maintenance, TS1: Cleaning and surface treatments.)
2.06 *Vegetation—is creeper concealing the true state of the stonework? Is vegetation rooted in decayed joints especially at the top of boundary and parapet walls?* The roots of plant life in the stonework will break down the mortar and force the stones apart, allowing water to penetrate, increasing root growth and causing rapid disintegration.
2.07 *Is moisture getting into the stonework; if so where does it enter?* Almost every defect that can occur in stonework is caused by the presence of water: rainwater carrying sulphates in solution from polluted atmospheres, rusting iron cramps and other ironwork expanding and causing spalling of the stone, **2**, saturation of porous stones which disintegrate by freezing. Moisture may enter through faulty flashings, gutters and rainwater outlets or through open joints in the stonework. Rising dampness may bring salts in solution from the earth which crystallise causing rapid decay at the base of a wall, especially if the ground level has risen as is often the case for old buildings.

Condition of joints
2.08 *Have joints perished or decayed? Are stones loose or in danger of falling? Is immediate action needed to ensure public safety,* **3***?* Faulty joints cause stones to be loose, allow rainwater penetration and encourage vegetation growth and frost action especially in exposed positions. Disintegration of stonemasonry walls can be rapid when joints are defective, irrespective of the condition of the stones.

Conditions of stones
2.09 *Are stones decayed, eroded or spalling? Is trouble extensive or patchy? If patchy, is the cause localised such as at a rainwater outlet? Does it affect only some features such as window dressings or quoins,* **4***?* The cause of defects may be obvious to the architect, for example overall decay caused by the use of an unsuitable stone or an occasional stone decayed through being face bedded. In some cases expert advice will be needed. Accurate diagnosis is of paramount importance only when it influences repair policies of replacing stones or of treating the stonework to retard decay.

3

4

2 *Window of entrance hall of Cobham Hall, Kent, where rusting of iron cramps or dowels has caused spalling.*
3 *Ely Cathedral: pinnacles seriously decayed. For the sake of appearance and public safety repair was essential.*
4 *Reynold's Chapel, Norwich. A large proportion of stones affected by spalling and the remainder likely to be affected soon.*

2.10 *What is the condition of any carved work or sculptures? Has it antiquarian or artistic value?* Carved work is by its nature and position often more exposed than other stonework. It is usually of relatively soft stone. Expert advice may be needed regarding antiquarian value, **5**.

5 *Heads on south front of Southwell Minster left to go into 'glorious decay', though corbels renewed.*

3 Defining the repair policy

3.01 The architect should be now have a clear idea of the problems, if not all the answers. One overriding factor in defining a policy will be finance—what is available or what may be raised by an appeal, for example. In the long term there is no doubt that gradual refurbishing is the most economical and sympathetic way of caring for old buildings. All too often, though, buildings get into an alarming state of disrepair and even danger before action is taken. Urgent restorations are carried out and the building is then neglected until the need for repair is again acute. Clients should, where possible, be persuaded to take a long term view of continuing maintenance.

4 Repair of general defects

4.01 No two jobs of stonework repair will be identical. To make this description of practical work easy to follow, it is presented in the sequence used for diagnosis.

4.02 *Structural defects*, which may require stitching of the cracks, consolidating the core of ancient walls or underpinning, should be dealt with before general stonework repairs. Careful diagnosis is crucial and may require accurate measurements over a fairly long period, **6**. If structural movement has finished, some simple stitching or rebonding will probably be adequate.

4.03 Thick walls of ancient buildings will probably comprise masonry skins with a rubble core which may be unstable. Cracking may be in one skin only or through the wall; or the core may settle, causing the wall to bulge. Part of the outer skin may need to be removed to consolidate the core, but in less serious cases grouting will provide sufficient stability. Where one skin is removed, stones should be marked and replaced in their original positions.

4.04 Weak walls or walls leaning at the top may be held together by an rc beam along the top within the core. The beam may be covered by rubble or a coping. For high walls beams can be inserted at various heights by removing short lengths of facework and fixing the beams into the core. Such beams avoid the need for buttresses and can be inserted as stitches across cracks.

4.05 Bulging in one leaf only may occur in seventeenth or eighteenth century buildings where walls were often of two unbonded skins with no core. Higher load on one skin will cause bulging in the other leaf, rectified by reconstruction or grouting. From the late eighteenth century the stone facing was usually bonded to a brickwork backing.

4.06 *Dirt and soot* should be cleaned off as required before repairs and repointing; for water cleaning, crevices and seriously opened joints should be stopped to prevent undue water penetration into the fabric. Cleaning reveals many defects, the true state of the joints and the real colour of the stone for matching. Ideally, cleaning should be done before writing the repairs specification.

4.07 *Creeper and vegetation* should be removed. Vegetation should be killed with weedkiller before removal from joints. Creeper should be cut off at ground level to allow it to die and weaken its hold.

4.08 *Moisture* from faulty flashings, gutters and rainwater outlets and rising damp causing decay in the base of a wall should obviously be put right. Water will also enter through copings, string courses and other projections if joints are open or weathering is insufficient. Open joints should be filled; the top surface of such projections are usually best covered. If the top is uneven it should be levelled with mortar and coated with bitumen to prevent corrosion of the underside of the flashing. The flashing can be lead or other suitable material, but not copper as this stains the stonework beneath.

6 *Settlement cracks tagged to check movement in Jewel Tower, Westminster (1948).*

7 The character of stonework is greatly affected by the type of pointing.

5 Joints: repointing

5.01 Pointing is not only important in maintaining stonework but good pointing will unify and improve facades composed of varying types of stone, **7**. Unfortunately workmen of sufficient skill are in short supply.

Raking out
5.02 Defective mortar and disfiguring recent pointing should be raked out to 20 to 40 mm and washed out to remove dust and reduce suction during repointing. Modern cement pointing is usually too hard to be raked out and may need the use of small compressed air tools, carborundum discs or chain saws. Modern cement pointing is not only unsightly but too strong, and when it shrinks water can enter by capillary attraction, especially on exposed walls.

Mortar for pointing
5.03 Mix varies with stone porosity, hardness and degree of exposure. It is usually either lime and sand, known as 'coarse stuff', or lime, sand and a little cement. Sieved stone dust sometimes replaces the sand.

Lime
5.04 Hydraulic lime is not readily available. France is the only current source. Non-hydraulic or semi-hydraulic lime is available as dry hydrate and occasionally as lump for slaking. The hydrate should be soaked overnight to form lime putty and kept moist till used. Non-hydraulic lime may be bought as lime putty. Hydrated hydraulic lime should not be soaked, as it will harden.

Sand
5.05 Sand should be clean and sharp and chosen to match existing pointing: some experimentation may be needed. For medieval work and rubble walls very coarse grit or crushed gravel is most suitable. A fine sand is needed for fine ashlar.

Cement
5.06 The use of a little ordinary Portland cement will speed hardening and increase strength and weather resistance. A little cement will normally be needed for non-hydraulic and semi-hydraulic lime mortars.

Colourants
5.07 Mineral dyes may be used if sand, lime and cement do not give the desired colour. Ready-mixed lime/sand mixtures are available ready pigmented for gauging on site. The colour is normally constant from batch to batch.

Mixes
5.08 The mortar should not be stronger than the stone. Lime/sand mixes of about 1:3 are average. These set slowly and should be kept moist for several days with sacking in dry weather. Lime/cement/sand mixes should have similar proportions of cementitious matter (cement + lime) to sand and the proportion of cement to lime should vary from 1:3 for less durable stone to 1:2 for tougher stone and 1:1 for tougher stones in exposed places such as copings, parapets and chimney stacks. An appropriate mix, containing lime, provides flexibility: plasticisers should not be necessary.

Type of joint
5.09 Ashlar walls and new stonework should normally have flush joints. Weatherstruck pointing is usually out of character with stonework.

Medieval walls
5.10 These walls were no doubt flush pointed originally, but arrises are now rounded and flush pointing would result in feathered edges and thicker joints. The feather edge breaks away in time, forming a lodgement for water and accelerating decay. So mortar must be set in to avoid feather edges and reveal the shape of each stone.

Finishing the joint
5.11 For wide joints with coarse aggregate mortar the joints should be stippled with a bristle brush to roughen the surface or sprayed with a hose to wash away the laitance. Stippling has the advantage of tightening the joint.

Galetting
5.12 Wide joints are vulnerable to weathering but the exposed area can be reduced by galetting—setting small chips of flint or ironstone into the joints while still wet.

6 Defective stones

Replacement
6.01 If stones are defective they should be dealt with before any repointing is carried out. The most straightforward treatment is to cut out decayed or damaged blocks and replace them, **8**. Some attention should be paid to the cause of the deterioration, as this may affect the choice of the replacement stone, especially if rising damp or incompatibility of stones have played a part. The total visual effect of renewal should always be considered. One of the major problems is deciding the extent of the renewal necessary. A total refacing is very unusual, and the replacement of partially deteriorated stones will relate to their probable life and their intrinsic antiquarian value, the visual effect of patching, the availability of suitable matching stone and the cost.

Where many stones are defective, and others likely to become so within a few years, it may be sensible to replace both types for the sake of appearance and long-term economy.

A time scale varying according to the height of the building, the relative cost of scaffolding and the regularity of maintenance or repair is useful. A return to carry out further restoration within 25 years would be uneconomical. Therefore it makes sense to try to assess the life of the stones which are being left undisturbed. A general guide on replacement is to leave all stones which have decayed back less than 50 mm, but this decision must be made sensibly to suit individual circumstances.

Replacement stone should, ideally, match the colour and characteristics of that which it is replacing, unless the deterioration of the original indicates a poor choice initially, in which case a visual match with better weathering characteristics should be sought.

Frequently substitutes have to be considered because

8 *Defective stones cut out to receive new ones on Southwell Minister.*

quarries are no longer working, or are producing stone from the same locality as the original, but with less durability. Advice on availability and matching may be obtained from the Secretary, Standing Joint Committee on Natural Stones,[1] or from the organisations listed in the Natural Stone Quarry Directory[2] and from the Natural Environment Research Council.[3] Local masonry contractors are often able to supply additional useful information. It is important to bear in mind that weathering characteristics are just as significant as appearance when selecting new stone.

Secondhand stone
6.02 There may be unwanted outbuildings or boundary walls that could be used, at least in part. The outer skin of double walls may be repaired with stone from the inner skin which in turn can be replaced by a substitute such as brickwork. Stones can sometimes be reversed if they are taken out carefully and have sufficient depth in bed.

Cutting out
6.03 Normally this is done carefully with hammer and chisel. Power saws may be used to open up fine joints but power hammers should not normally be used. Stones may be cut out to their full bed but more usually to 75 to 100 mm, and never less than 50 mm except for veneering small local patches (see paragraph 7.02).

Cutting and bedding
6.04 Much of the stone will be cut at the quarry or in the yard, but a good deal of cutting will be done on site in restoration work. The faces should be true, especially for fine-jointed ashlar. Machine cut faces will normally be true but the site cut faces need careful watching. Hollows or convexities may set up stresses in the stonework.

Stones should be laid on their natural bed except for cornices and copings which are best joint-bedded, and arch stones, which should be laid with their bed at right angles to the thrust. The natural bed is not always easily recognised and the block should be marked in the quarry to indicate bed.

Preserving essential characteristics
6.05 New stonework should match in every way, **9a, b**. Appearance will vary with the tool used and the craftsman's technique. Early Norman stones were cut with axes, later stones with chisels. It is generally inappropriate to replace them with stones finely sawn by machine. Stones should where possible be cut similarly to existing ones, for instance slightly rounded at the edges if existing stones are weathered. But artificial 'distressing' is not effective. Joints should match existing ones in colour, thickness, layout, etc. For example, in medieval work perpendicular joints are never over each other. In classical work they are much more regular. In coursed random rubble the stones will probably be hammer-dressed only and level beds will be 300 to 450 mm apart.

Colouring
6.06 Though it is anathema to purists, there may on occasions be no alternative to staining stone. Proprietary stains are available, though application requires some skill. Cowdung is sometimes used to speed weathering in rural buildings but is likely to introduce damaging salts.

9a

9b

9a, b *It is important to preserve the lines of mouldings, string courses, etc when replacing stone.*

Setting
6.07 Mortar mixes have already been dealt with. The wall should have bonding stones built into the backing at suitable intervals and if large areas of stonework are being replaced it is essential to put in new bonders at suitable (similar) spacing.

Anchorages
6.08 When mortar gives insufficient stability, dowels, cramps or other fixings are needed. Metal anchorages should be bronze alloy, stainless steel or galvanised iron or steel. Metal cramps are used to secure projecting cornices and string courses, gable copings and face stones of narrow bed. Dowels are used in mullions, columns, balusters, etc to prevent lateral movement on bed joints. Anchor bolts are used to secure projecting cornices into the backing or to a structural frame where there is insufficient holding-down masonry above (see also Processing, TS2: Site works and cladding).

Joggle joints and cement joggles
6.09 A joggle joint is a projection worked on one stone fitting into a matching recess on an adjoining stone. The joint is made loose enough to allow for some adjustment in setting and space for cement grout which is poured into the joint when the stones are in position. Cement joggles are formed by pouring cement into cuts, V-shaped in section and Y-shaped on elevation in adjacent sides of adjoining stones, **10**. When the stones are set, cement grout is poured into the channels formed by adjoining cuts. This restricts the individual forward movement of stones. Joggles are used mainly for cornice stones.

10 *Cornice stones at St Paul's Cathedral showing joggles.*

Carved work
6.10 It is often difficult to decide whether to renew badly eroded carved work. If the original form is clearly recognisable, usually no treatment is needed except replacement of chipped fragments such as fingers of statues. If, however, it is grossly deteriorated, a choice has to be made between continuing into 'glorious decay' and replacement, on the basis of cost, antiquarian value, availability of craftsmen and the information that exists of the original form. If the work is repetitive, less decayed stonework may be found elsewhere on the building which can be copied, or old photographs, sketches and other documentary evidence may be found. Where no evidence is forthcoming a client may accept, for example in a series of carved heads round a church, the replacement of the carved feature by an entirely new design.

6.11 If the carved work is fairly clearly detailed, it is a good idea to make a detailed photographic record while access is possible from a scaffold, both as an historical record and to inform future repairers.

7 Alternative repair methods

Piecing in
7.01 This involves renewing only part of the stone and is appropriate where spalling of part of a stone has been caused by rusting cramps or physical damage. Its use should be limited or the character of the stonework will suffer.

Veneering
7.02 Not normally recommended, it involves using very narrow stones 30-40 mm thick. The thinner the stone, the more anchorages are needed.

Redressing
7.03 Where weathering is uniform it may be possible to redress the whole face of a wall giving it a new lease of life. Difficulties occur at window and door openings. The method has been used for badly eroded cornices, recutting the mouldings to a slightly steeper angle, **11**.

Plastic stone
7.04 There is much controversy about its use, though within the limits of the material and used intelligently it can be effective and economical. Used by a competent craftsman and restricted to repair of small areas (at most single stones), the junction with original stonework is softer and less obtrusive than a replacement in natural stone, and considerably cheaper. It tends to look drab and dead in appearance compared with the quality and texture of natural stone, hence its restriction to small areas. Badly done, it can accelerate weathering of the adjacent natural stone.

7.05 Plastic stone is suitable only for cavities and losses between 25 and 75 mm deep. Decayed stone should be cut back to a firm base, preferably regular in shape and parallel to the original coursing. A key is formed by undercutting round small areas, dovetail fashion, so that the inside dimension is greater than the face. For larger areas dovetail shaped holes can be cut in the firm base into which plastic stone is forced, or copper (or other non-ferrous) wires about 3 mm diameter as reinforcement can be pebble plugged to sound stone. For cornices and heavy projections heavier reinforcement, securely fixed, will be needed.

7.06 Ingredients will vary according to colour, texture and density of stone to be matched. Mixes should be weaker than the stone; mixes as for pointing (see paragraph 5.08). Artificial colouring may be added to the mix but may reduce the strength of the repair. Colour will change with drying out so experimental samples should be made first. Sand is the usual aggregate, or crushed stone with the dust sieved out.

11 *A deteriorated cornice (below attic) recut to avoid expensive replacement.*

Filling with plastic stone

7.07 After cutting out and washing to clean and reduce suction, filling should be in one or two coats depending on thickness, with no feather edges. The filling should be tamped round reinforcement, **12a, b**. Finished with a wooden float, it should be left slightly rougher than surrounding stone. To fill across a joint, make the joint and point later, do not merely score the plastic mix. Slow setting should be encouraged.

12 *A church at Stone, Bucks, where* **a** *the friable stone has been hacked off and reinforcement placed, and* **b** *the plastic stone repair is completed.*

Failures of plastic stone

7.08 Many repairs have failed because:
- repairs are too large, resulting in shrinkage cracks, crazing and lifeless appearance;
- too dense mix causes cracking, crazing and rapid deterioration of adjacent stone;
- reinforcement using iron nails which rust and expand disintegrate the repair and allow moisture to enter the cavity accelerating decay;
- violent colour contrast by the use of wrong aggregate or stain produces an unpleasant patchy effect.

Epoxy and polyester resin repairs

7.09 Epoxy resin adhesives are very useful for refixing broken stones which have suffered accidental damage or spalling, **13a, b**. They are also used to repair small local defects by veneering.

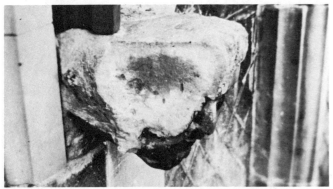

13a *A damaged carved head at Carlisle Cathedral, which was repaired by securing a new piece of stone with epoxy resin; and* **b** *carving in situ.*

7.10 Gothic window tracery can be successfully and economically repaired with the aid of epoxy resins. The technique has been used on Ely and Carlisle Cathedrals. In the past, when the face of window tracery decayed it was necessary to remove the glass and take out stones to their full depth and entirely rebuild the tracery. It is now possible to leave the glass and inner face of the tracery intact, hacking off the outer face level, up to the glass line if necessary, dependent on the extent of decay, **14**. New stones are cut and fixed with epoxy resin adhesive and non-ferrous dowels, **15**. Low viscosity resins may be used in grouting small fractures.

8 Specification

8.01 The architect cannot leave to the mason decisions about the jobs described here. A check list of items to be considered will be given with the specification clauses in section 6 of the handbook.

14

14 *Defective tracery to Lady chapel of Ely Cathedral hacked off to glass line prior to veneering with epoxy resin and new tracery.*
15 *View of the repaired tracery at Ely.*

References

1 Standing Joint Committee on Natural Stones, Admin House, Market Square (North Side), Leighton Buzzard, Bedfordshire.
2 Natural Stone Quarry Directory, Ealing Publications, 73A High Street, Maidenhead, Berkshire, SL6 1JX.
3 Natural Environment Research Council, Institute of Geological Sciences, Exhibition Road, London SW7 2DE.

Maintenance Technical study 3

Recent developments in in-depth treatments

The shallowness of their effect considerably limits the success of the cleaning and surface treatments covered in study 1 of this section. Recent developments have involved attempts to deal with stone in depth. M. J. BOWLEY describes experiments in poulticing to remove salts and C. A. PRICE describes attempts to impregnate stone in depth.*

Poulticing

1 Introduction

1.01 Surface treatments such as lime washes and silicone water repellants have generally been unsuccessful in preserving soluble salt-laden stonework. Observations reveal that these treatments have had little effect in arresting decay; in some instances enhanced decay has been suspected.[1]

1.02 Soluble salts migrate to the surface of wet stonework under drying conditions, then dry and crystallise causing decay. Experimental attempts are now being made to remove these salts by the in-depth method of poulticing as suggested in an early BRE digest.[2] The building used in the experiment described here is the Salt Tower, Tower of London.

Salt tower

1.03 The ground floor of the tower is a thirteenth century room with arched embrasures, walls of Kentish Rag coursed rubble and arches of Reigate ashlar quoins and voussoirs, **1, 2**. The Reigate stone is now considerably decayed, **3**.

1.04 As its name suggests, the tower was used for storing culinary salt when this was a highly valued commodity. Though the salt concentration is high it is not uniquely so and results of the experiments have considerable relevance elsewhere.

* The work described in this paper forms part of the research programme of the Building Research Establishment and is published by permission of the Director.

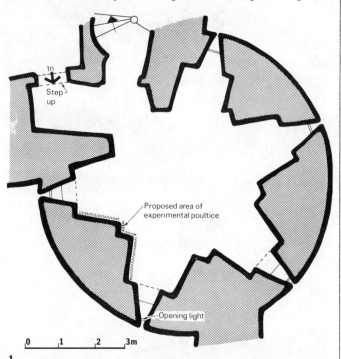

1 *Plan of salt tower showing massive stonework and area poulticed.*
2 *Interior view of stonework at junction of two embrasures in the salt tower.*
3 *Decayed Reigate stone quoins. The lower block is a Victorian renovation, the upper an original piece. Assuming the repair was inserted flush with the faces of the original, 20-25 mm of decay has taken place and the replacement is now itself beginning to decay.*

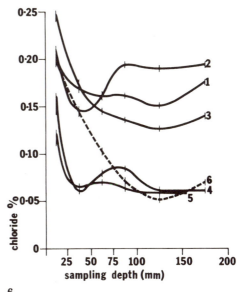
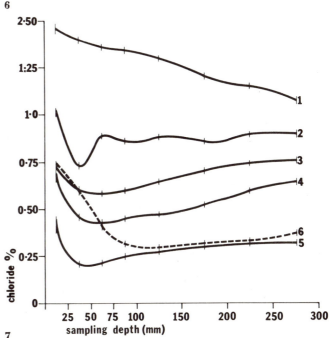

2 Poulticing method

2.01 Since soluble salts migrate in stonework only if water is present the stone must be wet for poulticing to work. Fine mist sprays are effective and economic; less than 200 litres of water per hour are sufficient to feed six sprays covering about 7 m². The sprays are set up to give, by visual inspection, a reasonable overall wetting. This continues for 7 to 14 days dependent on the type and porosity of stone, thickness of walls and their construction. Full experimental details are available from BRE,[3] **4**.

The poultice
2.02 The chosen absorbent, usually a clay or diatomaceous earth, is added to water and mixed until a consistency. rather like soft ice cream is obtained. The paste is applied to the wetted wall in a layer about 20-25 mm thick, **4**, and is held in place by embedded wire mesh. As the poultice dries from the outside it draws salt-laden water from the stone.

Duration
2.03 The poultice is allowed to remain for at least one month. Once it is hacked off the cycle of wetting and poulticing can be repeated as often as required. This is restricted by the need to protect the outside poultices from rain and the wetted walls from frost action.

Measurement
2.04 More than surface measurement is required for salts

4 *Applying poultice to stonework. The test area was up to 2 m high. The guttering was for removing the water run-off from wetting.*
5 *Collection of stone samples. A rotary-percussive drill is used for speed.*
6, 7 *Variation in distribution of chloride in depth, in blocks of 2 and 3 respectively. Curve 1 shows initial concentration. Curves 2-5 show concentrations after the four successive poultice applications. Curve 6 shows concentration one year after fourth poultice. Note that sampling depth of block 2 is 200 mm whereas on block 3 it is 300 mm. Sampling in 25 mm, then 50 mm lengths.*

and by far the simplest method of measurement through the thickness of the block is drilling, **5**, collecting the drilled-out stone at successive increments of depth and analysing in the laboratory for chloride content. Clearly this defaces the stone and plastic repair is required or cutting out and replacement of the sample block.

3 Results of poulticing

3.01 The two blocks used for sampling, shown at positions 2 and 3 in 2, have very different initial salt concentrations, indicated by the vertical axes in **6** and **7**.

3.02 As can be seen from these graphs the concentration falls fairly steadily throughout the stone with successive poultices, block 2, **6**, stabilising around 0·07 per cent. This concentration cannot be assumed to be low enough to stop further decay. Curves **6** and **7** show an increase in concentration one year after the fourth poultice, at least at depths up to 200 mm. This subsequent salt migration is accounted for by the lack of dpc and the dampness of the site, which is situated just above river level. The concentration of salt in the footings and subsoil is unknown. Isolation or removal of any reservoir of salt in footings or subsoil is an essential part of any remedial work.

Deep impregnation
4 Introduction

4.01 Stone preservation has had a long and disappointing history. No reliable inconspicuous surface coating has been found to prevent deterioration of stonework. Preservative treatments are urgently needed, **8**, and recent research has been based on impregnating stone in depth.

Stone decay
4.02 The major mechanisms of decay in the British climate dealt with in study 1 of this section are: the repeated crystallisation of salts within the pores of the stone, the formation of ice crystals in the stone, chemical attack by acid rainwater and shear stresses generated by differential movement between sulphate-clogged surface skin and the underlying stone. Though more detailed information is available,[4, 5] it is sufficient to note here that the crystallisation of salts is the major cause of stone decay in Britain and that common sources of salts include soil water, sea spray, unsuitable cleaning materials and the reaction of limestone with acid rainwater, **9**.

Failures of surface treatment
4.03 There are two main reasons why shallow surface treatments may accelerate decay[1]: 1, water invariably gets behind the treated layers, but is prevented by the treatment from evaporating at the surface of the stone. Instead it has to evaporate from just behind the treated layer and any salts in solution crystallise there and may cause spalling; 2, the treated layer differs in moisture and thermal movement properties from the underlying stone and this may lead to shear failure. If treated layers could be made sufficiently thick, at least 25 mm rather than the current 2 or 3 mm, some of the problems associated with shallow treatments would be overcome. In-depth treatments would consolidate the stone and prevent further deterioration caused by salt crystallisation by encapsulating the salts in resin and/or making the stone more resistant to damage. Such an increase in resistance could involve an increase in the stone's tensile strength and/or a modification of pore structure.

5 Impregnation methods
Suitable materials
5.01 A liquid is required that will penetrate deeply into stonework and then solidify. For adequate speed of penetration, its viscosity must not be much more than that of water, **10**. It should not set too slowly as it would then tend to evaporate or spread too thinly through the stone.

8 *Severe decay in Doulting limestone, Wells Cathedral.*
9 *Acidic rainwater and salt crystallisation have caused this Portland stone carving to weather into holes.*
10 *The impregnant has to be applied intermittently over a period of several hours to achieve adequate penetration.*

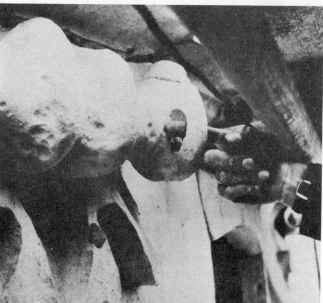

11 This statue of Doulting limestone has been impregnated with an alkoxysilane by conservators at the Victoria and Albert Museum.

12 An emperor's head from the Sheldonian Theatre, Oxford, after impregnation by immersion in molten wax.

Too fast setting on the other hand would prevent penetration.

5.02 Toxicity, flammability, volatility, water miscibility, flexibility after curing and cost must all be considered, but the viscosity requirement is the more stringent. It is fulfilled either by dissolving a resin in a low viscosity solvent or by using liquids consisting of small molecular units (monomers) which subsequently react within the stone to give a large molecular network (a polymer).

Solvents
5.03 There is a danger that resin will migrate back to the surface as the solvent evaporates. Also large resin molecules may be too big to enter the smallest pores. The solvent occupies pore space during impregnation that would otherwise be filled with resin. Nevertheless, Gauri claims to achieve successful penetration using an epoxy resin dissolved in solvent and this is marketed in the USA.[6]

Alkoxysilanes
5.04 Of the monomers which polymerise in situ, alkoxysilanes have received most attention. Some of them are directly available from chemical manufacturers, others are obtained in various mixtures as proprietary stone preservatives, **11**. Curing is dependent on reaction with water and loss of alcohol by evaporation. This means that it is impossible to achieve complete pore filling and the treated stone remains permeable to water vapour.

Acrylics
5.05 These monomers are not usually practicable for use on buildings since they have to be cured by heating. Objects which can be removed can be treated in the laboratory. Munnikendam has developed an acrylic system which cures at room temperature but curing is inhibited by oxygen so he has not been able to use the method effectively for large objects.[7] A further disadvantage is that curing commences on mixing so that viscosity increases during application and inhibits penetration.

Low molecular weight epoxy resins
5.06 Unlike common epoxy resins these monomers are of low viscosity. Though a reaction between the hardener and carbon dioxide in the air produces white efflorescence this can with care be avoided.

Molten wax
5.07 This is usually impracticable for impregnation of stone in situ because of the need to heat the stone to a temperature higher than the melting point of wax. However, it can be used on small objects, **12**, and is sometimes used by museums for sculpture.

Limewater
5.08 This is a suspension of lime in water. It should not under any circumstances be used on sandstone since lime plus sulphur dioxide in the air form calcium sulphate which may cause rapid decay. Its usefulness is debatable in application to limestone. Instances are known where hardening of the stone and an increase in its useful life have apparently been achieved, but attempts to reproduce this effect have rarely succeeded. There are insufficient grounds to recommend its widespread use.

6 Cost

6.01 The amount of material required depends on many factors such as porosity and depth of impregnation but as an example 5 litres/m^2 are required to fill the pore space 25 mm deep in a stone of 20 per cent porosity.

6.02 Assuming bulk purchase the prices per litre of some of the more important materials are: paraffin wax 15p, one type of acrylic 35p, alkoxysilanes ranging from 90p to £2·30. It is evident that it would normally be unrealistic to contemplate the treatment of large areas of plain walling at these material-plus-labour costs but costs do not prohibit the conservation of valuable details.

7 Conclusions

7.01 Little can be said as yet about the effectiveness of impregnation treatments since there is no reliable test for the durability of the treatment. RILEM (Réunion Internationale de Laboratoires d'Essais et de Recherches sur les Matériaux et les Constructions) and BRE are currently working on this. Natural exposure trials are unsatisfactory because they are very slow although they do serve to identify any treatments which might accelerate decay. Accelerated decay has not so far been attributed to any of the impregnation treatments described here. This, added to the theoretical justification for deep impregnation and the prospects of laboratory tests for durability of treatment, gives rise to optimism that deterioration of irreplaceable stonework may soon largely be avoidable.

References

1 B. L. Clarke and J. Ashurst, *Stone preservation experiments*, Building Research Establishment, 1972.
2 Building Research Station *Digest* (1st series) no 20, 1950.
3 M. J. Bowley, Building Research Station Current paper CP 46/75.
4 *The decay and conservation of stone masonry*, BRE *Digest* no 177, May 1975.
5 R. J. Schaffer, *The weathering of natural building stones*, HMSO, 1932 (reprinted BRE 1972).
6 K. L. Gauri et al, *Studies in conservation*, vol 18 p 25, 1973.
7 R. A. Munnikendam, *Studies in conservation*, vol 12 p 158, 1967.

Specification Technical study 1

Testing porous building stone

In specifying stonework the architect may require stones to be of particular characteristics. The wide range of tests available for dealing with porous stones is described here by C. A. PRICE, with the emphasis on durability.* Information on testing laboratories can be obtained from BRS, the quarry or stonemason, or from Stone Industries (see TS2 appendix 1 for addresses).

1 Strength

1.01 Strength tests are sometimes necessary to size loadbearing members. Even the weakest will bear some imposed loads but abnormally high stresses causing failure may result from such causes as uneven settlement or constrained thermal expansion.

1.02 The standard engineering tests are used on specimens of standardised moisture content, ie after immersion in water for 24 hours, **1**. Though specimens may be wetter and thus weaker than they would become in most conditions of use, it is the plinth courses and column bases which carry the greatest loads and it is these features that become wettest during flooding.

1.03 The strength of stone does not provide any indication of its resistance to weathering.

2 Chemical tests

2.01 Detailed chemical analysis of stone is rarely worthwhile. The chemical composition of a limestone, for example, is no guide to its durability. Even with sandstones, where durability is largely dependent on the chemical nature of the cementitious material that binds the grains together, general chemical analysis may be uninformative since the results cannot indicate which constituent represents the binding material.

2.02 A less comprehensive test involving immersion in 20 per cent sulphuric acid (specific gravity 1·145) is more informative. Sandstones in which the binding material is not resistant to acid lose their cohesion in this test and must be regarded as having low durability when used externally, **2**. Those unaffected after 10 days may be regarded as resistant to direct attack by acidic constituents in the air and in rainwater, but the test gives no indication of the stone's resistance to crystallisation damage or to contour scaling.

2.03 Immersion in acid also serves to distinguish between limestones and sandstones, although this distinction can often be made more readily by scratching the surface with a penknife; limestones are generally appreciably softer than sandstones. When immersed in acid, limestones dissolve rapidly with the evolution of carbon dioxide and little or no residue remains. Calcareous sandstones react more slowly to leave a crumbly mass of silica and sandstones with a resistant binding material[1] are unaffected.

3 Porosity

3.01 Porosity is defined as the volume of a stone's pore space expressed as a percentage of the stone's total volume. It is most conveniently measured by saturation with water under vacuum, though this does not take into account any closed pores within the stone which are inaccessible to water. The measurement of porosity is described in appendix I.

3.02 Knowledge of porosity is of limited use. It is required for the calculation of thermal transmittance (see Specification, IS 1: 'Thermal transmittance of stone'), for calculating the quantity of material needed to impregnate a particular stone to a given depth, and gives a rough guide to durability, though it is not of itself a sufficient measure. Stones of low porosity can generally be said to be more durable but the distribution of pores in terms of their size has greater bearing on durability than overall porosity.

*The work described has been carried out as part of the research programme of BRE (DOE) and is published by permission of the director.

1 *Compression test in progress.*

2a *A resistant sandstone and* **b,** *a calcareous sandstone at the end of an acid immersion test.*

Porosity is not very useful even in indicating the permeability of stone to water.

4 Saturation coefficient and water absorption

4.01 When porous stone is placed in water, the rate of absorption is initially very rapid but slows down after a few hours even though the pore space is only partially filled. Complete saturation may take many years. The amount of water absorbed in the short term by any given material is dependent on the soaking time and on the degree of immersion.

4.02 Various countries have adopted their own conventions. In Britain and a number of others water absorption is defined as the volume of water absorbed in 24 hours (expressed as a percentage of the total volume of the test piece) when the test piece is completely immersed in water. The saturation coefficient is the ratio of the water absorption to the porosity. It is in effect the proportion of the total pore space that is filled during the 24-hour soak. The measurement of water absorption and saturation coefficient is described in appendix II.

4.03 Saturation coefficient and water absorption are crude measures of pore structure. Stones with a high proportion of coarse pores have relatively low saturation coefficients, and vice versa. The value of the saturation coefficient can range from about 0·40 to 0·95, and at either end of the range the saturation coefficient provides an excellent indication of durability. A value of 0·50, for example, indicates a stone of high durability, whereas a value of 0·85 indicates low durability.

4.04 However, many building stones have saturation coefficients in the range 0·66 to 0·77 and in this region the saturation coefficient is an unreliable guide. Consideration of saturation coefficient and porosity together[1] provides a better indication, but this is still inferior to the results obtained from the crystallisation test (see para 8.01).

5 Microscopy

5.01 Microscopy is another technique permitting investigation of a stone's pore structure. A specimen of the stone is impregnated under vacuum with a coloured resin and a section of the stone approximately 0·025 mm thick is prepared. The distribution of the pore space can be examined and the constituent minerals identified. As the process is laborious and demands considerable skill, other techniques are generally to be preferred. Nevertheless microscopical examination may sometimes be useful for identifying a particular type of stone.

6 Pore size distribution and microporosity

6.01 Of the available techniques the two most widely used for determining the distribution of pore sizes are mercury porosimetry and the suction plate technique.[2] The underlying principle of both is that the pressure required to force mercury into an empty pore (or suck water out of a full pore) is dependent on the size of the pore. The test piece is therefore subjected to a range of pressures and the mercury (or water) content is measured at each pressure. The results may then be related to the pore size distribution of the stone.

6.02 In general the higher the proportion of fine pores contained by a stone, the lower is its durability. An 0·005 mm diameter has been found to represent a convenient dividing line between 'fine' and 'coarse' pores; the proportion of the total pore space that is accounted for by pores with an effective diameter less than 0·005 mm is termed the 'microporosity'.

Durability of limestone

6.03 Microporosity has been found to provide a useful indicator of the durability of many limestones (Clipsham being one of the main exceptions), but is of little value in predicting durability of sandstones. A microporosity of less than 30 per cent would indicate a limestone of exceptionally high durability that should give excellent service under any exposure conditions, whereas a value of 90 per cent or more would indicate a poor quality stone that should be used only in rural inland areas in situations protected from frost. In the 50-80 per cent region microporosity provides a less clearcut indication of durability, but here the combined use of microporosity and saturation coefficient has been found helpful.

Durability factor

6.04 Richardson has developed a technique based on water and air permeabilities which permits the measurement of a 'durability factor'[3, 4]. The durability factor is related to the proportion of coarse pores and the technique may therefore be expected to be applicable to those stones where microporosity is also a useful guide to durability. The particular advantages of Richardson's technique are that it can be used in situ and that it is rapid and thus cheap.

7 Freezing tests

7.01 Great difficulties have been encountered in developing a laboratory freezing test that reliably reproduces the known behaviour of building stones under natural conditions. An accelerated test employing natural frosts in which samples stand out of doors partially immersed in water has been found to give more reliable results. It suffers from its dependence upon natural frosts; a period of several years may be required to get significant results, **3**.

7.02 The formation of ice from water is essentially a crystallisation phenomenon. The following test which primarily assesses the resistance of stone to the crystallisation of salts has been found to give an indication of frost resistance too.

8 Sodium sulphate crystallisation test

8.01 The main cause of stone decay in the UK is the crystallisation of salts within the pores of stone. It is

3 *Tagged samples standing in water during a natural freezing test.*

4 *The typical state of samples of (left to right) excellent, good, moderate and poor durability after the sodium sulphate crystallisation test.*

therefore not surprising that the best single test for assessing the general weathering resistance of stone should be one in which samples of stone are subjected to cycles of immersion in a salt solution followed by drying in an oven, **4**. Sodium sulphate has been found to be the best salt for this purpose. The sodium sulphate crystallisation test is described in appendix III.

8.02 The detailed experimental procedure must be adhered to exactly. Recent collaborative tests with European laboratories have shown that seemingly minor variations of the procedure can lead to complete reversal of the results; the reasons for this are being investigated. The only permissible variation is the substitution of saturated sodium sulphate solution for 14 per cent sodium sulphate solution as the salt. The modified test is more applicable to sandstones than to limestones and is used when the sandstone will be required to withstand particularly severe exposure conditions or if a particularly long life is required.

8.03 The crystallisation test is a comparative rather than an absolute measure of durability. Samples of stone of known durability under conditions of intended exposure must therefore be included with the test samples. Limestone should always be used for comparison with limestone; sandstone with sandstone. The main disadvantage of the crystallisation test is that it takes several weeks to complete. It is consequently rather expensive and cannot be used for day-to-day quality control of a quarry's output.

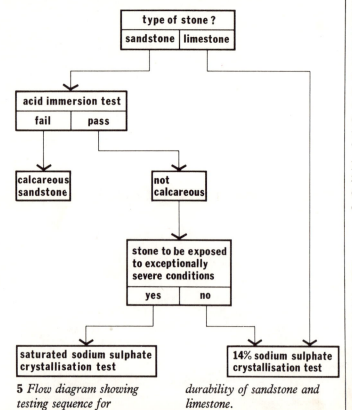

5 *Flow diagram showing testing sequence for durability of sandstone and limestone.*

9 Conclusion

9.01 The main emphasis of this study has been on the assessment of durability. The flow diagram, **5**, shows the BRE procedure for assessing the durability of an unknown type of stone.

Appendix I: Measurement of porosity

Dried samples are placed in a vessel connected to a vacuum pump and fitted with a tap funnel to admit water. After evacuating to low pressure (less than 3 mm of mercury) and maintaining this for two hours, air-free water is run in until the samples are well covered. Air is then re-admitted. After at least 16 hours immersion the samples are weighed suspended in water (W_1), weighed in air after removing excess moisture from the surface with a moist cloth (W_2) and then dried in an oven at 103°C. They are then cooled and reweighed (W_0). The porosity is given by $100 (W_2-W_0)/(W_2-W_1)$ per cent.

Appendix II: Measurement of water absorption and saturation coefficient

Measurements are made as described in appendix I. The dry samples are then immersed in water at room temperature and atmospheric pressure for 24 hours, wiped with a moist cloth and weighed (W_3). The water absorption is given by $100 (W_3-W_0)/(W_2-W_1)$ per cent. The saturation coefficient is given by $(W_3-W_0)/(W_2-W_0)$.

Appendix III: Sodium sulphate crystallisation test

A representative number of cubes of 40 mm or 50 mm side are cut from the stone under test and from stones of known weathering behaviour. They are dried in an oven at $103 \pm 2°C$, and then immersed for two hours in separate beakers of sodium sulphate decahydrate (specific gravity 1·055 at 20°C). The temperature of the solution is thermostatically controlled at $20 \pm 0·5°C$. They are then dried in an oven arranged to provide a high relative humidity in the early stages of drying and to raise the temperature of the cubes to $103 \pm 2°C$ in not less than 10 and not more than 15 hours. This can be achieved by placing a tray of water in the cold oven and switching on half an hour before putting the cubes in; 300 ml of water is adequate in an oven capable of taking 48 cubes. The cubes are left in the oven for at least 16 hours and are then cooled to room temperature before resoaking in a fresh sodium sulphate solution. This cycle is carried out 15 times (except when the cubes break up before this). The cubes are weighed after the final drying if they are sufficiently coherent. The results are reported in terms of the loss of weight expressed as a percentage of the initial dry weight, or as the number of cycles to failure if the cubes fail in less than 15 cycles. Conclusions are drawn by comparing the results obtained with the stone of known weathering behaviour.

References

1 D. B. Honeyborne and P. B. Harris, *Proceedings of the 10th Symposium of the Colston Research Society*, Butterworths, London, 1958.
2 Road Research Technical paper 24, HMSO, 1952.
3 B. A. Richardson, *Stone Industries*, September/October 1967, p16.
4 B. A. Richardson, *Stone Industries*, March/April 1971, p22.

Specification Technical study 2

Notes and specification

Specifying stone for building requires considerable knowledge of the properties of local stones, so the architect must rely on local observation of stone in use and expertise of the quarry manager in selecting an appropriate stone. JOHN ASHURST indicates what knowledge of stone the architect needs to have and provides specification notes for stone on site. REG WOOD provides a checklist of items to be considered when preparing a specification for repairs.

1 Introduction

1.01 Selection of stone for building should be based on personal observation and on knowledge of sources and characteristics of available stones. Most areas have a characteristic stone which may be local or which, by frequent use, has become familiar in a foreign environment. Considerable justification is required for introducing a 'new' stone into a setting where it may forever seem out of place.

1.02 Local observation will show how particular stones may be expected to weather, and in what situations they should be handled with special care or avoided altogether.

Availability
1.03 Full inquiries should be made on what stones are available. *The Stone Directory*[1] of the Stone Industries is invaluable in this respect but it is not enough to establish that a quarry is still working. The directory cannot hope to be completely comprehensive or to keep pace with the changing situation, even though it is revised and updated every two years. Potential supplies, current block or slab sizes and colour range should all be studied at the earliest possible stage with the help of the quarry managers.

1.04 Further advice may be sought from many sources, especially the Federation of Stone Industries, the Association of Natural Stone Industries (NFBTE), the Institute of Geological Sciences (especially on identification and origin, and possible matching), and the Building Research Establishment (DOE) (see appendix I).

Selection
1.05 The quarry manager's advice is essential on the selection of 'good' block. Where large supplies are envisaged over a considerable period of time there may even be justification for the consultant services of a *quarry inspector*, whose brief would be to seek out suitable supplies, liaise with the quarry manager and make periodic visits to both the quarry and the site during the work to ensure that the agreed quality and colour are maintained. In normal circumstances, however, this service is reliably provided by the quarry manager in direct liaison with the architect.

1.06 Certain defects are visually apparent after only a limited experience of quarry inspection and much can be learnt from a competent inspector on the job. Fine vents and faint traces of soft beds will soon be detected. A sound block freed from the face will ring like a bell when struck with a hammer or piece of stone whereas a fissured block will sound dully. However, the architect should not carry out selection in the quarry unaided without considerable practical knowledge.

1.07 Quarry inspection should start at the pre-contract stage. If necessary the specification should limit the stone to be used to certain specific quarries, and even specified beds where considerable variations may be present in quality and/or colour in closely associated strata. Visits during progress should check on consistency of the material and ensure that this as well as the workmanship is in accordance with the specification.

Testing
1.08 There are usually sufficient data available on the stones in common use to provide the necessary evidence on likely durability. Retesting of such stones is rarely carried out unless for some specific reason such as suspected contamination or sudden deterioration in quality. However, tests may be desirable on unfamiliar stones or stones which will be subjected to extreme conditions, or where weathering evidence suggests inconsistency in performance.

1.09 Of the tests available, those determining strength to assist the calculation of loadbearing members, those determining the distribution of pore sizes and microporosity and those determining the resistance to salt crystallisation giving the best guide to durability, are of most importance (see Specification, TS1: 'Testing porous building stone'). Frequently experience and observed performance of similar stones are in themselves the most valuable guide.

2 Specification notes

References
2.01 Any specification for stone in building should be closely based on the new BSI Code of Practice for Stone Masonry (formerly CP 121.201 and CP 121.202) BS 5390:1976. Specification (The Architectural Press) and the National Building Specification are also recommended for guidance.

Selection
2.02 The quarry and bed/s should be specified as matching the range of colour, texture and quality of samples already supplied and approved.

Drawings
2.03 The supplier should provide drawings prepared from the architect's detail drawings showing all dimensions, key numbers and fixings, all of which should be checked and approved before any work is put in hand.

General
2.04 Masonry walls should be constructed in accordance with the recommendations of BS 5390:1976. Variations on

natural bedding should be specified. Beds, face and back to be worked square and true from end to end, and stones to be of the full dimensions for bonding on face. Specify bedding and pointing in one operation unless special pointing is required.

Finishes
2.05 Specify finishes to be in accordance with agreed samples (sawn, rubbed, tooled for ashlar, a variety of hammer dressings for rubble).

Thicknesses and tolerances
2.06 Refer to thicknesses shown on drawings. For granite, limestone and slate, stones 50 mm thick, or less, should not vary more than 1·5 mm in 900 mm over height or length, and stones over 50 mm not more than 3 mm in 900 mm. The thickness of rubbed or frame-sawn slabs should be within 3 mm of the specified thickness, although riven-face slabs can vary up to 9 mm.
The length and height of slate slabs should be within 1 mm of specified sizes.
Marble slabs should be worked to within 1 mm of specified sizes, and their thickness to within 3 mm.
12 mm clearance should be left between backing and cladding.

Fixings
2.07 Specify metal wall ties to be of copper, phosphor bronze or stainless steel to comply with BS 1243.
Tying back and linking fixings should be copper, bronze or austenitic stainless steels.
Load-bearing fixings should be of aluminium bronze, phosphor bronze or austenitic stainless steel.
(Austenitic stainless steels from sheet, strip and plate to BS 1449 Pt 4, and from bar, to BS 970 Pt 4).

Mortars
2.08 Refer to BS 890 for non-hydraulic lime (hydrated and putty) and specify that hydrated lime run to putty should stand for 12-16 hours before use. If a hydraulic lime is to be used (in the UK this will be an imported lime, since hydraulic lime is no longer available) this must not be soaked and precautions must be taken to store it dry and ensure that the bags are not broken.
Ungauged lime mortars, containing no cement, should not be used in situations where high early strength is required.
Cements should comply with BS 12 (1971), BS 4027 (1972), BS 146 (1958), and BS 4248.
Masonry cement should comply with BS 5224.
Sand should comply with BS 1200.
Pigments should comply with BS 1014.
Pozzolanic materials such as pulverised fuel ashes should be sampled to ensure that there are not unacceptable levels of soluble salts present.

Typical mixes are:
Lime: sand   2:5   ⎱ suitable for most building
Cement: lime: sand 1:3:12 ⎰ stones
Cement: lime: sand 1:2: 9 suitable for exposed details
Cement: lime: sand 1:1: 6 suitable for most sandstones
Cement: sand   1:3   suitable only for dense granite
Pozzolanic PFA: lime: sand 1:1:4 suitable for less durable stones in sheltered environment

In freezing conditions, work must either be postponed or properly protected and if necessary heated. Air-entraining plasticisers rather than chloride 'anti-freezers' should be incorporated to improve frost resistance, provided that the resultant reduction in density and up to 10%-15% loss of strength is acceptable. Care in mix specification is especially important when bedding and pointing stone. The strength and porosity of the stone should be matched as closely as possible by those of the mortar.

Damp proof courses
2.09 Materials should comply with BS 743 and should be bedded in mortar not weaker than 1:¼:3 (cement:lime:sand). Suitable materials are code 4 (2 mm) lead-cored bituminous sheet, bituminous felt, black polythene sheet not less than 0.5 mm thick, slates in two courses bedded in 1:3 mortar (cement:sand).

Joint thicknesses
2.10 All the stones should set on a full bed of mortar with all joints and joggles completely filled and all mortices and cramps properly grouted.

Specify the joint thickness, eg:
Fine work	2–3 mm
Internal cladding	1.5 mm (marble)
External cladding	2–3 mm
Slate cladding	3 mm
Large slabs	4.5 mm
Polished granites	4.5 mm
Fine ashlar max.	6 mm
Rubble walls	12–18 mm

Expansion and compression joints
2.11 These are to be formed where shown on the drawings. In loadbearing masonry these must be raked out and left open during construction, to be pointed and cleaned off on completion of the work. In cladding the joints must be squeezed out and jointed as work proceeds.
Specify building mastics to comply with BS 3712 and sealants to BS 4254. Manufacturers' recommendations must be followed on the appropriate back-up material. Appendix B in CP 121: Part I: 1973 summarises the properties of sealing compounds.

Scaffolding
2.12 Putlog or independent scaffold may need to be of the double type if extensive structural and stone repairs are to be carried out, and should consist of two lines of standards parallel with the building, linked to each other by transoms and ledgers and tied at openings. Aluminium alloy scaffolding should be recommended if the scaffold is to be in position for a long time. Ends of tubes should be capped to prevent damage. All scaffolds must comply with statutory regulations and should be in accordance with CP 97: Parts 1, 2 and 3.

Storage
2.13 Stones should be stored clear of the ground and kept covered in winter. A special gantry should be constructed for the ashlars and adequate protection against damage on site provided. All stones should be checked for damage on arrival on site, and stored in sequence for fixing.

Protection and cleaning down
2.14 The general contractor is responsible for the above protection against weather, damage and staining by other trades. Timbers, oil, wet straw and washings from steelwork are common problems to be guarded against. On completion of the work, the masonry contractor should thoroughly clean down the work and attend to any defects in the pointing/polishing, etc., before handing over to the general contractor, who will then assume sole responsibility for the condition of the work.
Limestone which will be subjected to long periods of exposure to building operations should be given a protective slurry of lime/plaster which must be thoroughly cleaned off as the scaffold is struck.

3 Checklist for repair

3.01 This is a checklist of items to be considered in preparing a specification for repair. These are additional to the usual preliminary clauses, common to almost every building specification, such as form of contract, by-laws, notices, insurance, etc.

General
3.02 Importance of building historically/architecturally; special features; purpose of present contract; site workshop and off-loading facilities; dimensions to be taken from building.

Structural
3.03 Defects requiring attention; provisional sums for repair, underpinning, stitching, grouting, etc; replacement of fallen stones (ruins); marking stones taken down for refixing in identical positions.

Cleaning
3.04 Pc sum (specialist subcontract); or method described.

Creeper, etc
3.05 Removal (kill first).

Moisture
3.06 Flashings, gutters, rainwater disposal; cornices, etc, covering; rising damp treatments: pc sum (specialist subcontract), dry area, lower soil around walls, etc.

Joints
3.07 Raking out; mortar: materials, mix; type of joint; finish (texture); sample pointing for approval; no pointing in freezing weather.

Defective stones
3.08 Extent of renewal, architect's direction; match existing, all respects; cutting out (depth, power tools—restrictive, support work where large areas removed, remove iron cramps, removal of defective stone from site); stone for replacement (source, quarry, bed/second-hand sources); extent of working stone on site/at works; accuracy; tooling; setting (natural bed unless directed otherwise, bed to be marked at quarry—sedimentary, no voids in joints, binding stones, anchorages, joggles, etc); mortar for bedding (materials, mix); pointing as joints (para 3.07).

Carved work
3.09 Protection of: repairs; replacement (recarving in workshop—remove and transport, recarving in situ—facilities for carvers); stone for carving, quarry, bed; provisional sum or pc for carving.

Plastic stone repairs
3.10 Extent, architect's direction; competent craftsmen only; cutting out for plastic repair: depth, undercutting; anchorages, reinforcement; mix—less dense than stone; materials; colour matching: sand, pigments; texture; samples for approval; filling (wetting); removal of laitence.

Resin adhesives
3.11 Extent of use, by architect's direction; precautions in use (rubber gloves, hygiene), protection of surrounding work from runs.

Reference
1 For a list of currently producing quarries and their stone see *Natural stone directory* 1974, Ealing publications, 70 Chiswick High Road, London W4.

Specification Information sheet 1

Thermal transmittance

The thermal transmittance (U-value) of stonework should be calculated with careful attention to exposure, moisture content and porosity. C. A. PRICE describes the calculation procedure and some underlying principles and provides an example of the procedure in use*.

Introduction

The use of U-values
1.01 The thermal transmittance (or U-value) of an external wall is a measure of the wall's ability to conduct heat in or out of the building. It is generally used in calculating the flow of heat out of a building; the greater the U-value the greater the heat loss.

Limitations of current methods
1.02 BS CP 121.201[1] and 121.202[2] contain nomograms for determining the thermal transmittance of various types of wall without any knowledge of the transmittance of each component part. The transmittance of the combined construction can be simply read off, saving calculation. However, these nomograms are unsatisfactory because they are applicable only to a limited range of wall constructions and dimensions and they fail to take account of the moisture content of the wall.

General expression
1.03 These deficiencies can be overcome using the more general expression for the thermal transmittance, U:

$$U = \frac{1}{R_{si} + R_{so} + R_{cav} + R_1 + R_2 + \ldots}$$

R_{si} and R_{so} represent the inside surface and outside surface resistances, R_{cav} is the resistance of the cavity (if any) and R_1, R_2, etc are the thermal resistances of each of the solid components, **1**. However, the expression is not valid for surfaces involving heat bridging (ie a resistance is not constant for the whole wall); the IHVE guide[3] and BC CP3 chapter II[4] describe procedures for dealing with this and other more complex situations.

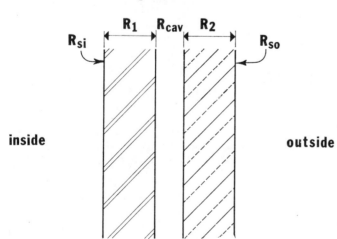

1 *Example of disposition of thermal resistances of a cavity wall.*

2 *Guarded hot plate apparatus for measuring thermal conductivities to BS 874.*
3 *Wall laboratory for measuring the thermal transmittance of various wall constructions. The interior of the laboratory is maintained at constant temperature, and the flow of heat through the walls is measured. This enables the values of the surface resistances R_{si} and R_{so} to be determined.*

Direct measurement
1.04 Though direct measurement, **2, 3, 4**, might seem an obvious alternative to calculation, it rarely gives the same results. This is because results obtained by direct measurement depend on conditions during tests and may differ among themselves, let alone from the calculated values in which assumptions about exposure conditions are implicit. Moreover, the actual U-value of a particular structure does vary to some extent from one situation to another dependent on such factors as wind speed and internal conditions. Difficulties may arise when regulations require that the thermal transmittance should

*The work described in this paper forms part of the research programme of the Building Research Establishment and is published by permission of the Director.

4 An alternative method for measuring the thermal transmittance of a wall. The 'hot box' which is fixed to the wall is maintained at constant temperature.

not exceed a given value or when it is necessary to compare different constructions on a common basis. To overcome these difficulties the adoption of standardised U-values has been recommended by IHVE and BRE.

2 Data for calculation

Standardised U-values

2.01 Standardised U-values are based on assumptions about the rates of heat transfer to and from surfaces by convection and radiation. These assumptions determine the values of the inside and outside surface resistances, R_{si} and R_{so}, and the cavity resistance, R_{cav}. The figures in table I for surface resistance are for high emissivity materials such as stone and brick and the cavity figures are for an unventilated air space or a cavity in a hollow wall construction. If the structure has two unventilated cavities such as a cavity wall with a batten-mounted dry lining, the expression for U-value should incorporate the sum of the appropriate resistances for each of the cavities. In the absence of a cavity, $R_{cav} = 0$. Usually thermal resistance is expressed in $m^2 °C/W$ and U-value in $W/m^2 °C$.

Table I Standard surface and cavity resistances				
	R_{si}	R_{so}	R_{cav} (cavity 5-20 mm)	R_{cav} (cavity 20+ mm)
Thermal resistance $m^2 °C/W$	0.123	0.055	0.11	0.18

Exposure

2.02 The standard value given in table I for outer surface resistance, R_{so}, implies normal exposure conditions. Non-standard values of R_{so} for other exposure conditions should be selected from table II. The three exposure conditions should be interpreted as:
sheltered—up to the third floor in city centres;
normal—most suburban and country buildings; fourth to eighth floors in city centres;
severe—buildings on the coast or exposed on hill sites; floors above the fifth of buildings in suburban or country districts; ninth floor and above of buildings in city centres.

Table II Variation of outside surface resistance R_{so} with exposure			
Exposure conditions	Sheltered	Normal	Severe
R_{so} ($m^2 °C/W$)	0.08	0.055	0.03

Resistance of solid components

2.03 The thermal resistance of a solid component, R_1, R_2, etc is derived from its thickness, l, and thermal conductivity, k.

$$R = \frac{l}{k}$$

If resistance is expressed in $m^2 °C/W$, l should be in metres and k in $W/m °C$. Measured conductivity values should be used when available; otherwise values for granite, marble and slate can be taken from table III. These are typical values and considerable deviation from them may occur in practice. However, it is unlikely that this will lead to serious errors in calculated U-values since granite, marble and slate are usually used as thin slabs which will make only a small contribution to the total U-value of the building shell.

Table III Generalised values of thermal conductivity			
Type of stone	Granite	Marble	Slate
Thermal conductivity k ($W/m °C$)	2.5	2.0	2.0

Conductivity and porosity

2.04 The thermal conductivity of limestone and sandstone is strongly dependent on its porosity. The test to determine porosity was described in TS1, 'Testing porous building stone'. The few data available have been used to construct the graph, 5, which should be used to estimate the thermal conductivity of a sandstone or limestone in the absence of direct experimental results.

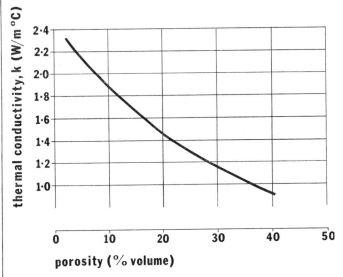

5 Graph of variation of thermal conductivity with porosity for limestone and sandstone.

Conductivity and moisture content

2.05 The thermal conductivity of stone is also dependent on its moisture content. The moisture content of granite, marble and slate will be very low under all circumstances and the data of table III can be used without further correction. For limestone or sandstone, however, the conductivity derived from the graph, 5, should be multiplied by the appropriate factor given in table IV.

2.06 Moisture content should be measured where possible since prediction is made difficult by variations of stone type, of exposure conditions, of effectiveness of any dpc and of moisture content across the thickness of the wall. However, if no more reliable data exist the moisture content of limestones and sandstones should be taken as 5 per cent where the stonework is exposed to rain and 1 per cent where it is not.

Using available data

2.07 Table V gives thermal conductivities for concrete and brickwork, materials likely to be used in conjunction with stone. The moisture factors of table IV may be used to adjust these conductivities to other moisture contents but it must be noted that the figures in table V already contain a moisture factor.

2.08 For example, for a bulk dry density of 200 kg/m³ and a moisture content of 5 per cent the conductivity is 0·12 W/m °C (table V). The figure for 10 per cent can be obtained by first dividing by the 5 per cent moisture factor, 1·75 (table IV), and then multiplying by the 10 per cent factor, 2·1.

2.09 Insulating materials are intended for use in dry situations and their conductivity in air-dry conditions is therefore appropriate.[3, 5]

Table IV Moisture factor for variation in moisture content

Moisture content (% by volume)	1	3	5	10	15	20	25
Moisture factor		1·3	1·6	1·75	2·1	2·35	2·55	2·75

Table V Thermal conductivities for concrete and brickwork

	Thermal conductivity W/m °C		
Bulk dry density kg/m³	Brickwork protected from rain: 1%*	Concrete protected from rain: 3%*	Brickwork or concrete exposed to rain: 5%*
200	0·09	0·11	0·12
400	0·12	0·15	0·16
600	0·15	0·19	0·20
800	0·19	0·23	0·26
1000	0·24	0·30	0·33
1200	0·31	0·38	0·42
1400	0·42	0·51	0·57
1600	0·54	0·66	0·73
1800	0·71	0·87	0·96
2000	0·92	1·13	1·24
2200	1·18	1·45	1·60
2400	1·49	1·83	2·00

* Moisture content expressed as a percentage by volume

3 An example

3.01 In calculating the standard U-value of a cavity wall with an outer leaf of limestone 150 mm thick and an inner leaf of brickwork 105 mm thick, the bulk dry density of the brickwork will be taken as 1600 kg/m³ and the porosity of the limestone as 20 per cent. The expression for U-value is:

$$U = \frac{1}{R_{si} + R_{so} + R_{cav} + R_1 + R_2}$$

where R_1 represents the resistance of the brickwork and R_2 the resistance of the limestone.

3.02 In this instance $R_{si} = 0 \cdot 123$ m² °C/W, $R_{so} = 0 \cdot 055$ m² °C/W and $R_{cav} = 0 \cdot 18$ m² °C/W (table I).

$R_1 = \frac{l_1}{k_1}$ where l_1, the brickwork thickness is $0 \cdot 105$ m and k_1 its thermal conductivity is $0 \cdot 54$ W/m °C (see table V).

$R_2 = \frac{l_2}{k_2}$ where l_2, the stone thickness is $0 \cdot 15$ m and k_2 its thermal conductivity is $1 \cdot 46$ W/m °C (see graph, **5**) multiplied by the moisture factor for 5 per cent, $1 \cdot 75$ (table IV) since it will be exposed to the rain (see para 2.06). Inserting these figures in the U-value expression gives:

$$U = \frac{1}{0 \cdot 123 + 0 \cdot 055 + 0 \cdot 18 + (0 \cdot 105/0 \cdot 54) + (0 \cdot 15/(1 \cdot 46 \times 1 \cdot 75))}$$

$= 1 \cdot 6$ W/m² °C

3.03 This example illustrates the contribution of each component to the total U-value. Thus the resistance of the limestone $(0 \cdot 15/(1 \cdot 46 \times 1 \cdot 75) = 0 \cdot 059$ m² °C/W) is less than one-third of that of the brickwork $(0 \cdot 105/0 \cdot 54 = 0 \cdot 194$ m² °C/W) even though the limestone is almost one and a half times as thick. This shows that it would not be worth while attempting to reduce the U-value of the wall by increasing the thickness of the limestone. Similarly it can be seen that the resistance of the brickwork is only slightly greater than that of the cavity. One way to achieve a substantial reduction in U-value would therefore be to create a further cavity by installing an internal wall lining mounted on battens.

Calculating heat loss

3.04 The heat loss through any particular part of a building shell may be obtained by multiplying together the area of that part, its U-value, and the temperature difference between inside and outside.

Table VI gives the thermal conductivity values of some common insulating materials which may be associated with masonry walls or cladding, as an aid to calculating the latters' U-values.

Table VI Thermal conductivity values for insulating materials

Material Bulk dry density kg/m³	Thermal conductivity k (W/m °C)
Asbestos cement sheet	0.36
Asbestos insulating board	0.12
Corkboard	0.042
Expanded polystyrene	0.033
Fibreboard	0.050
Glass fibre	0.035
Plaster, Gypsum, 1120	0.38
Gypsum, 1280	0.46
Perlite, 400	0.079
Perlite, 610	0.19
Sand, Cement	0.53
Sand, Cement, Lime	0.48
Plasterboard, Gypsum	0.16
Perlite	0.18

References

1 *Masonry walls, ashlared with natural stone or with cast stone*, BS CP121.201, 1951.
2 *Masonry-rubble walls*, BS CP121.202, 1951.
3 IHVE guide, *Book A, Design data*, IHVE, 1970.
4 *Thermal insulation, in relation to the control of environment*, BS CP3, chapter II, 1970.
5 *AJ Handbook of building environment*, section 4, Design guide table VII.

Stone in use Detail sheet 1

Bonded ashlar walls

Ashlar is defined as masonry consisting of blocks of stone, finely dressed to given dimensions and laid in courses with thin joints. When ashlar is bonded to a brick (or concrete) backing the construction is referred to as a traditional ashlared masonry wall. This bonded wall is to be found in many existing buildings but is not commonly used today. Modern use of ashlar relies on metal fixings as shown on detail sheets 7 and 8.

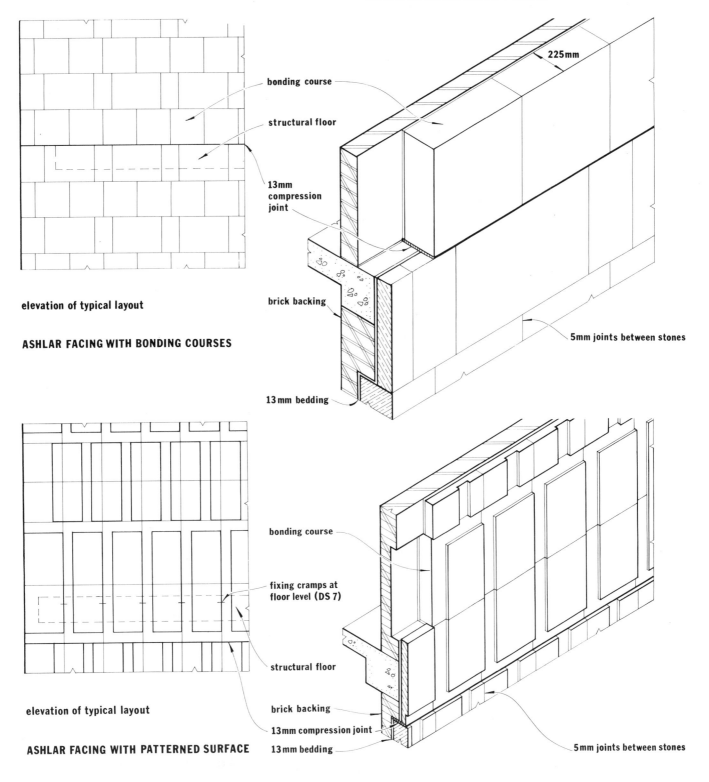

elevation of typical layout

ASHLAR FACING WITH BONDING COURSES

elevation of typical layout

ASHLAR FACING WITH PATTERNED SURFACE

Stone in use Detail sheet 2

Solid stone walls and windows

The solid stone wall, like bonded ashlar masonry, is to be found in many existing buildings, though in modern practice to meet building requirements it would normally be lined internally with insulating blockwork or used in a cavity wall construction. Thus detail sheet 2 shows examples of existing construction that the architect may come across during repair or reconstruction work.

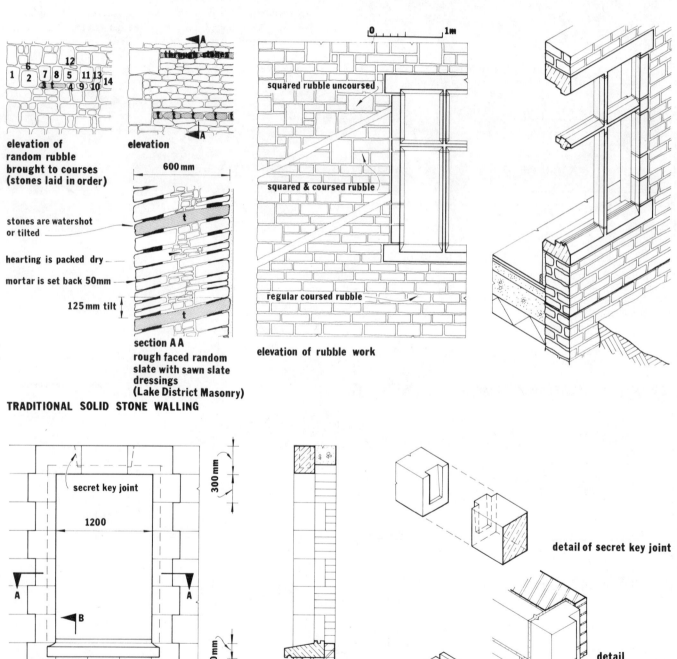

TRADITIONAL SOLID STONE WALLING

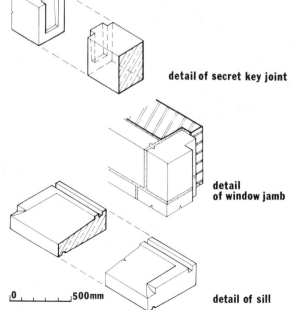

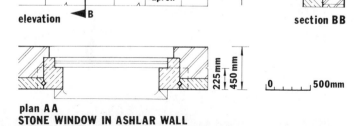

STONE WINDOW IN ASHLAR WALL

Stone in use Detail sheet 3

Rubble walling

Rubble work, unlike ashlar work, is primarily a local craft where the locally quarried stone determines the type if not the actual style of the walling. The examples in this sheet show details of solid rubble walls or cavity walls where the external skin is built in rubble.

The Code of Practice 121.202, which outlines the recommendations for this form of construction, is published as one code of practice with 121.201.

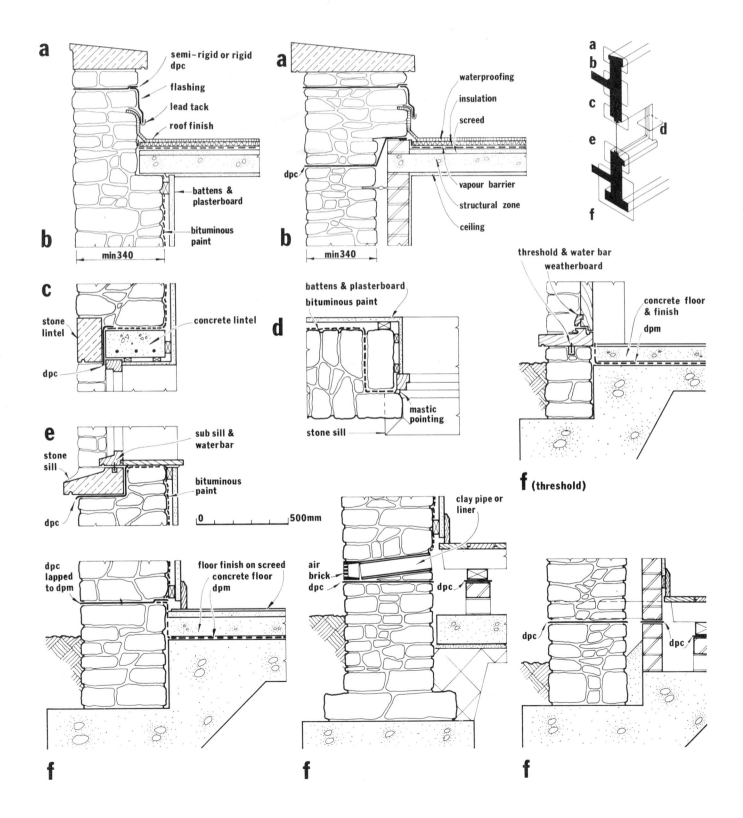

Stone in use Detail sheet 4

Masonry walls ashlared with stone

Detail sheets 4 and 5 consider typical details of bonded ashlar where the outer masonry wall is brick, ashlared with natural stone. The mortar normally contains a finer aggregate than that used in rubble and the stonework may be bedded and pointed with the same mortar at the same time.

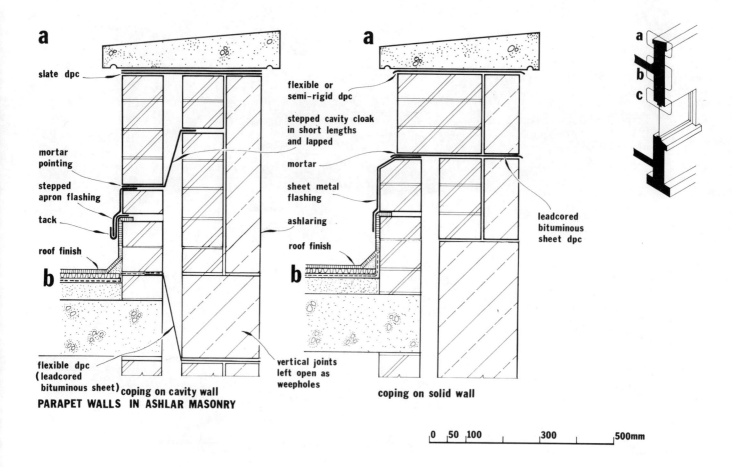

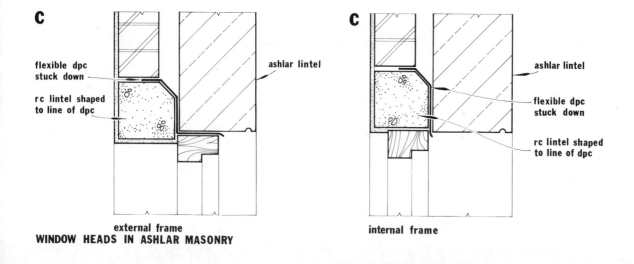

Stone in use Detail sheet 5

Masonry walls ashlared with stone

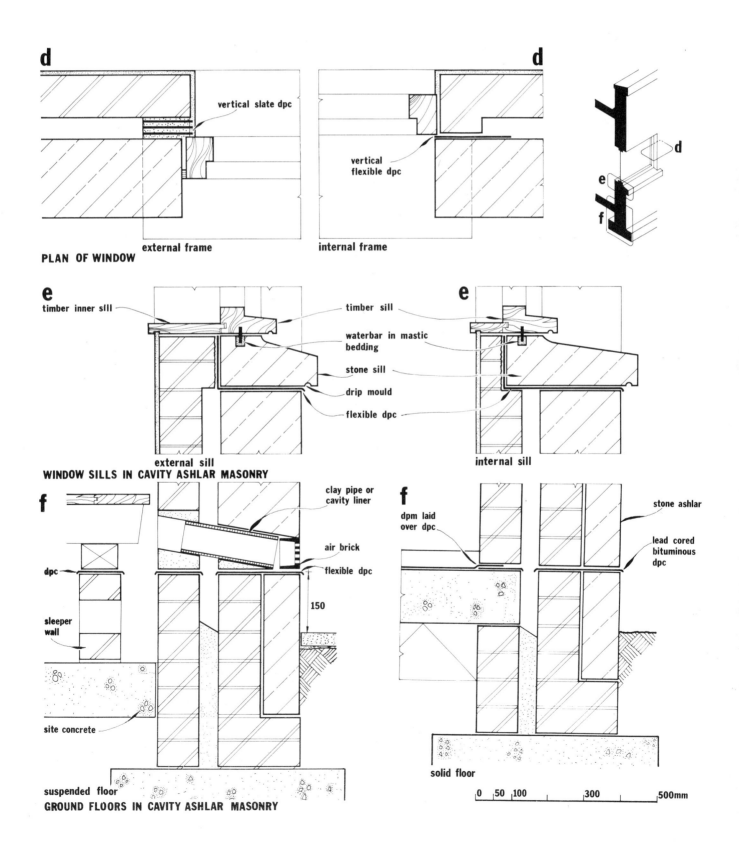

Stone in use Detail sheet 6

Stone faced concrete walls

As stone facings have become thinner and are now in the range of 75 to 100 mm, improved methods of fixing have been evolved for use in cladding. The fixings have two functions: 1 support of the dead load or loadbearing fixings (LF); and 2 withstanding the loads imposed by wind and structural movement or restraint fixings (RF). This detail sheet illustrates how the fixings are shown on elevation to indicate location of fixings.

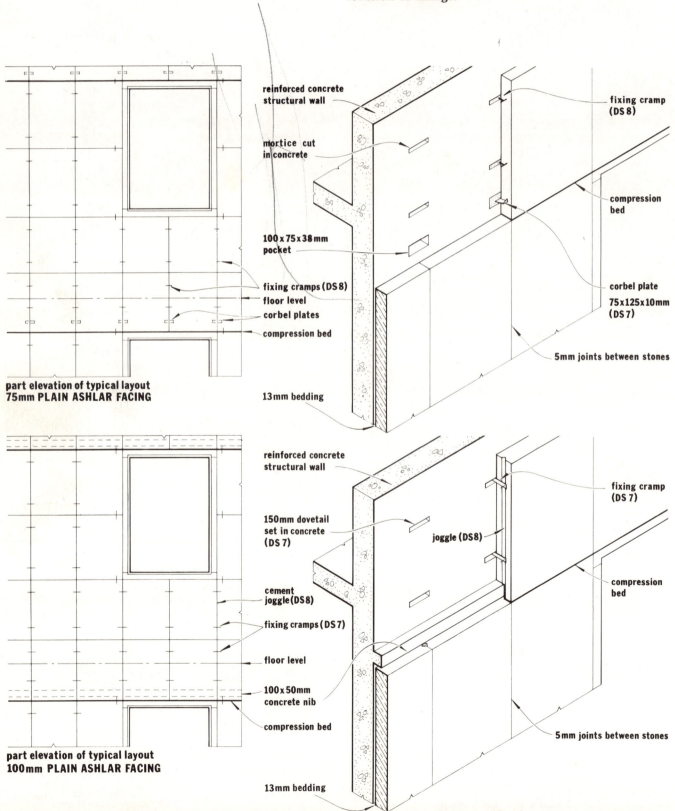

Stone in use Detail sheet 7

Fixings for ashlar facings over 75 mm thick

Materials for loadbearing fixings recommended by CP 121 (aluminium bronze, phosphor bronze, stainless steel) are used predominantly to fix the thicker ashlar of sedimentary stones where the dead weight of the stone itself is so great. Restraint is most necessary at the top and bottom edges of the upper and lower course respectively. The fixings shown illustrate to the architect the wide variety available, but the application relies on the individual choice of the mason on site.

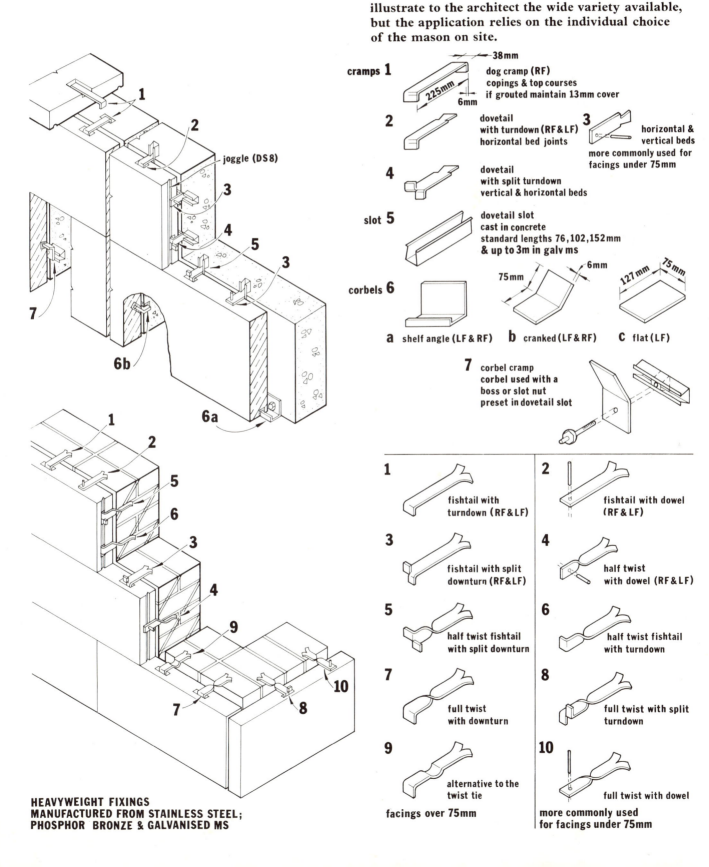

HEAVYWEIGHT FIXINGS
MANUFACTURED FROM STAINLESS STEEL;
PHOSPHOR BRONZE & GALVANISED MS

Stone in use Detail sheet 8

Joggle joints and fixings for ashlar facings under 75 mm thick

The joggle joint is a butt joint used between two ashlar stones with a cement dowel in the joggle. Materials for restraint fixings recommended by CP 121 are copper, phosphor bronze and stainless steels. These fixings apply particularly to the thinner facing which is weak in tension and this sheet illustrates the use of copper wire cramps.

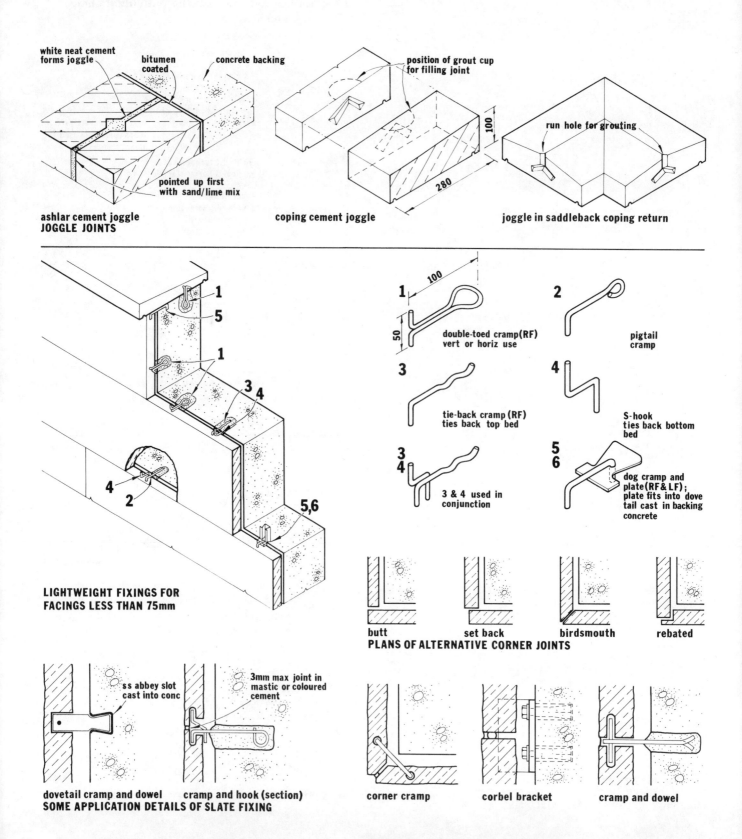

Stone in use Detail sheet 9

Copings and soffits

Parapet walls faced with stone are shown here with a variety of copings. The specification for natural stone copings can be found in BS 3798, section 5.
For further details of suitable compression joints see CP 298: 1972, figure 1.
The method of fixing stone soffits will depend on the depth of soffit required. A typical assembly is shown here where a single channel is satisfactory for soffits up to 300 mm wide.

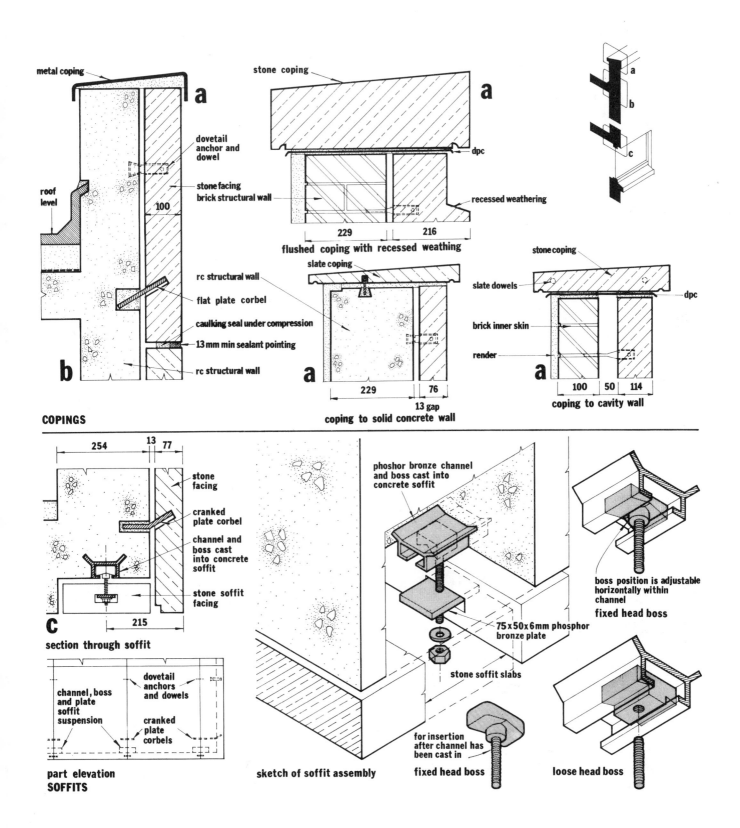

Stone in use Detail sheet 10

Slate roofing

The more exposed the position and the smaller the slate, the steeper the pitch must be: illustrated here are a few alternative methods of fixing with the different gauges required for diminishing or regular coursing.

Stone 'slating' is split thicker than true slates to allow for the greater permeability of the sedimentary stones. Rarely used for new work now, mostly repairs.

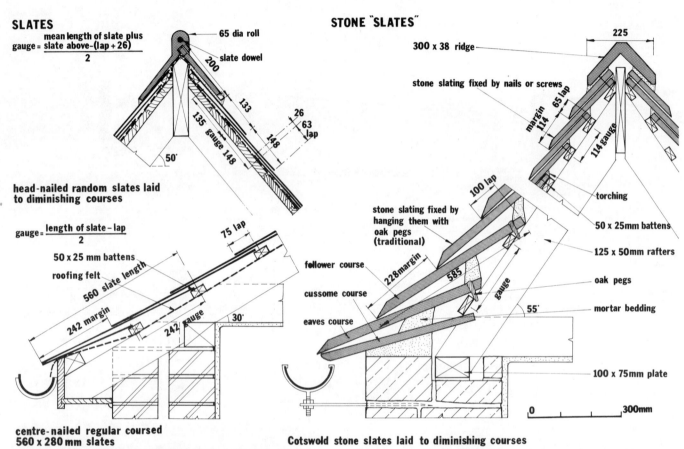

1 Bedding of slates where slater used a mortar mix of 1:4:16.
2 Capping off of a roof, using Dorset stone slates.
3 A slater's and plumber's paradise at Ashridge House, Berkhamsted.
4 A swept valley detail. Loose slates packed tightly to sweep valley and 'raise level'.

Stone in use Detail sheet 11

Stonework drawing

An illustration taken from the AJ of 24 January 1923, where Frederick Chatterton points out the merits of 'Architectural building construction' by Messrs W. R. Jaggard and F. E. Drury. In Chatterton's words, the illustration combines authentic practical data with well designed examples of their application.

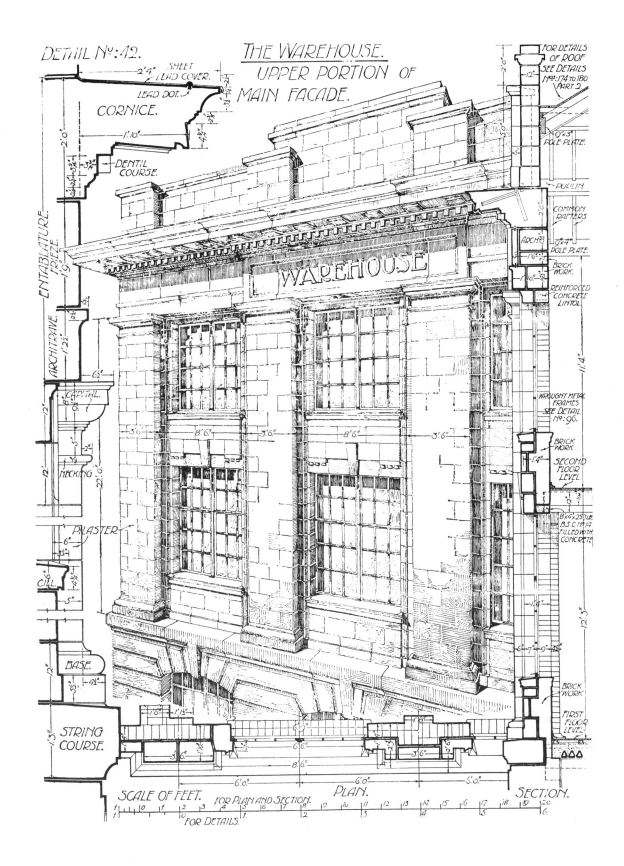

Stone in use Detail sheet 12

Stonework drawing

An illustration taken from 'Pencil Points' (an American journal) in January 1924, showing part of the Madison Place elevation of the US Treasury, Washington DC by D. C. Cass Gilbert, architect.

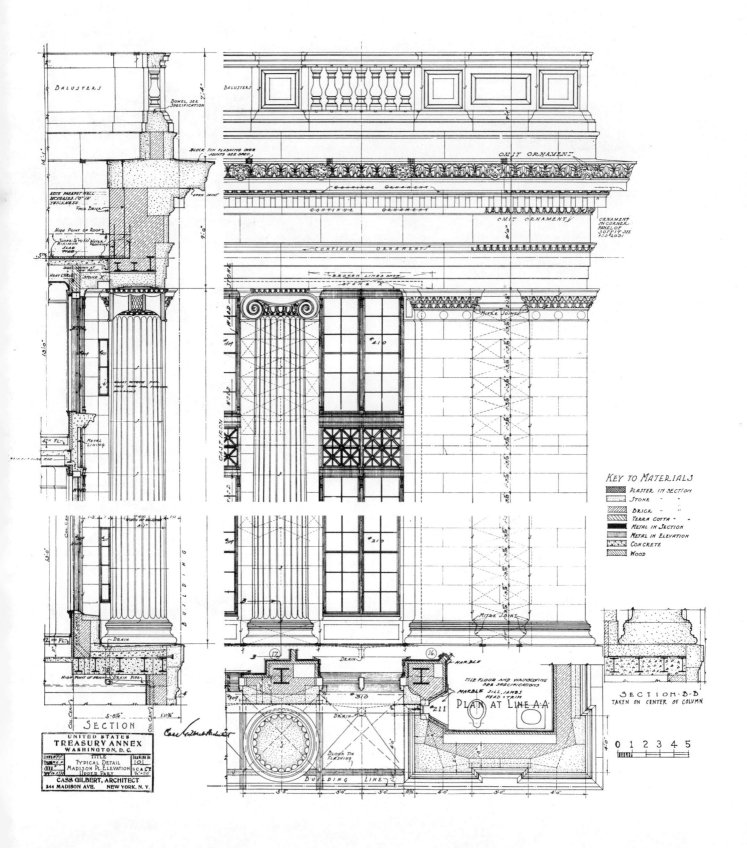

Stone in use Detail sheet 13

A modern stone building

The building illustrated is in Queen Victoria Street (London EC4) designed by Ansell & Bailey, architects; the stonework is fixed by Ashby & Horner (stonemasonry department).
It is not intended as a working drawing but as an illustration of the dimensions required to be shown. As every stone is numbered and cut off site, accuracy of measurement is of paramount importance.

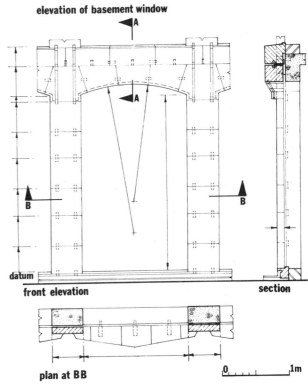

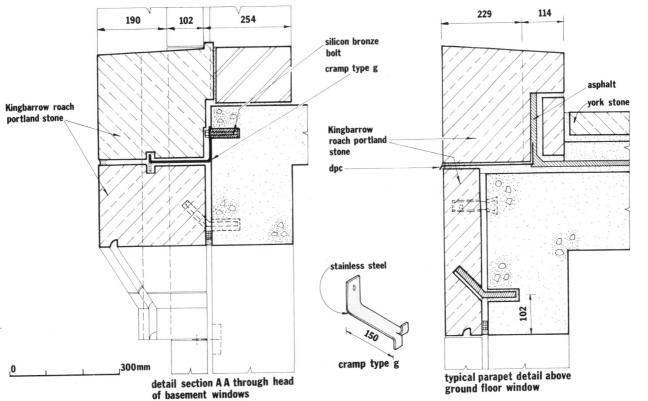

Stone in use Detail sheet 14

A traditional stone building

A stone portico has been built in Portland stone on the north face of 39 Old Steine, Brighton, by David A. P. Brookbank & Partners, architects of Haywards Heath. The work has been carried out by Stone Firms Ltd. Careful attention has to be paid to detail in this kind of work.

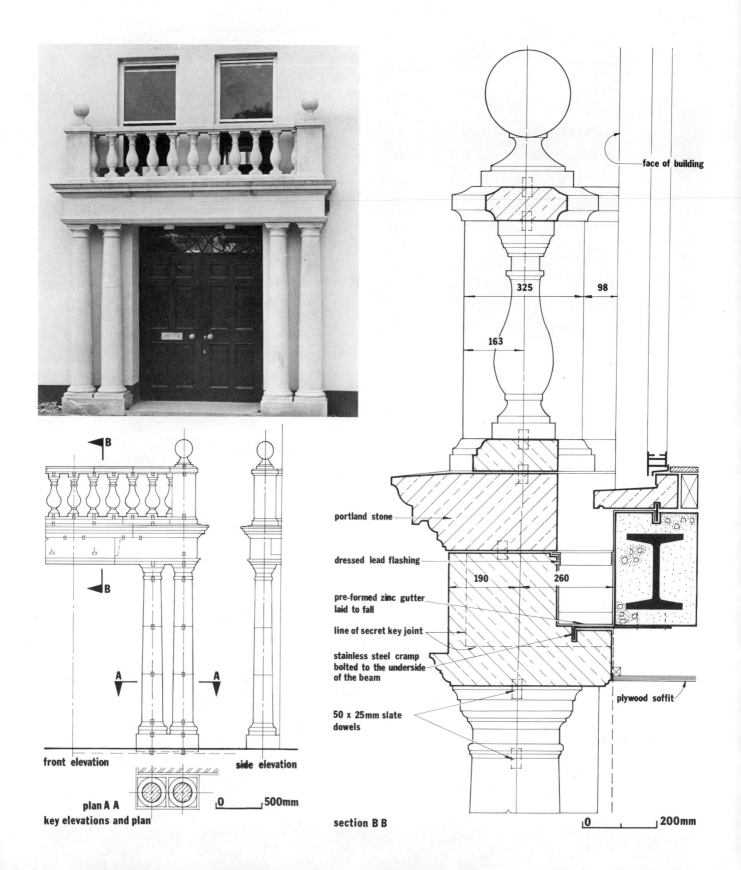

Stone in use Detail sheet 15

Buildings in stone

Three buildings of different scale and type are illustrated here; stone has been used as a facing material or, in one case, as a solid wall construction.
1 Lancaster Divisional Police headquarters; architects: Lancashire County Council.
2 'Ansty Plum' private dwelling; architects: Alison & Peter Smithson.
3 Economist building; architects: Alison & Peter Smithson.

1

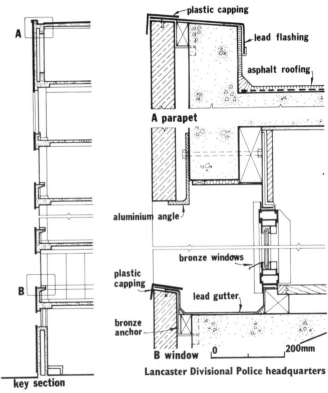

Lancaster Divisional Police headquarters

2

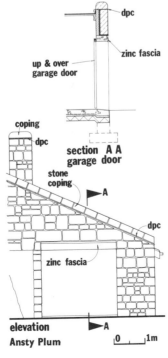

elevation Ansty Plum

3

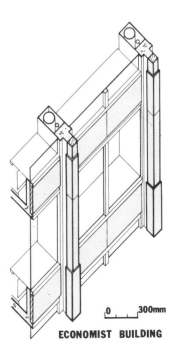

ECONOMIST BUILDING

Stone in use Detail sheet 16

Buildings in stone

A building by GMA International for Lewis Shopholdings in Brentham Halt Road, London W5, is illustrated here to show the combined use of marble and slate in thin wall cladding.
The travertine was fixed by J. Whitehead. The fixing of the slate relies on the strength of the adhesive. adhesive.

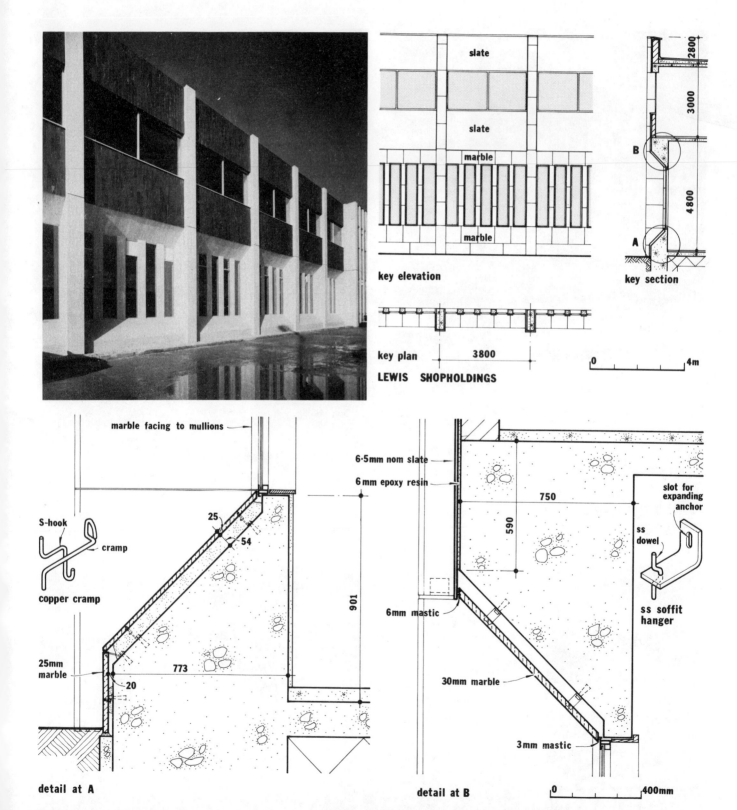

Stone in use Detail sheet 17

Decorative use of stone

A page illustrating the decorative use of stone that an architect might specify himself or commission from a sculptor.

1

1 *Relief (on the Time and Life building) by Henry Moore.*
2 *Clay model.*
3 *Finished stone head after carving, sanding and polishing.*
4 *Lazarus sculpture by Jacob Epstein.*
5 *Unicorn from the Sheldonian Theatre (Clipsham stone).*
6 *Restoration pediment for Queens College, Oxford.*

2

3

4

5

6

Stone in use Detail sheet 18

Decorative use of stone

A page illustrating the variety of uses of stone in decoration.

1 *The Carfax Conduit, Nuneham Courtenay, Oxford, carved in Headington limestone in 1617 and restored in 1959 and 1976.*

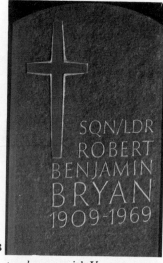

2 *Family crest in bathstone.*
3 *Westmorland greenslate tombstone with V-cut lettering.*

4 *Church font in Portland stone.*
5 *Halifax Building Society headquarters by Building Design Partnership.*
6 *Detail of battered wall.*

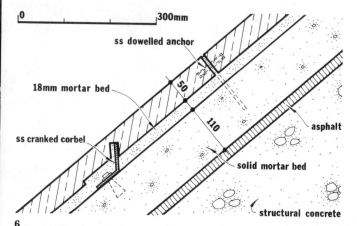

Technical study

Hard landscape in stone

Hard landscaping has been comprehensively discussed in two series: 'Concrete', and 'Brick' both published as books.* Much of the information contained in these studies applies equally to stone, and many of the examples already illustrated have included stone as part of the landscaping, demonstrating how successfully it may be used. This study has been written by JOHN ASHURST.

1 Selection

Of all landscaping materials stone is most at home in both rural and urban environments. This is not to say that any stone may be equally acceptable visually in any part of the country (regardless of environment). But few materials improve so reliably with weathering as well selected stone, and few are able so consistently to provide such a variety of texture, interest and dignity.

In the selection of stone for paving, boundary and retaining walls, and miscellaneous landscape features such as fountains, statuary and rockeries, consideration should be taken of the natural and planned environment of which it is to be a part, and of its durability, detailing, maintenance and cost.

Paving materials such as flagstones of carboniferous sandstone, granite kerbs, and setts and cobbles have been used so ubiquitously that they rarely seem out of place today. The same cannot be said of granite and marble plinths and retaining walls, especially in areas locally associated with limestone, sandstone or slate. Perhaps the most common example of a beautiful material uncomfortably misapplied is the white Italian marble found in almost every English churchyard, **1**. Local stone used locally will always look well; if it is to travel far then care must be taken. The walls and steps of millstone grit, and drystone masonry of carboniferous limestone, are in perfect harmony with both the house at Haddon Hall and the natural landscape of Derbyshire from which they have been quarried; but there may be some incongruity in the retaining walls and setts of granite alongside sandstone steps and paving, Clipsham limestone and Hampshire flint which have been used in the new buildings and landscaping of the Winchester law courts, **2**. Sometimes, however, there are good reasons for taking a material such as granite out of its natural environment, for it is extremely durable and is not normally vulnerable to frost damage and staining as are some other stones.

2 Durability

Paving

The attractions of natural stone paving still outweigh the cheaper cast slabs where prestige or high priority is given to landscape. There is ample evidence around the country of the long life of many stone pavings. Until a short time ago stone was the main material used for paving roads and footways, even after the introduction of tarmacadam and tarred wood blocks.

Pavements and large pedestrian circulation areas were commonly paved in 2in or 3in (50 mm or 75 mm) tooled or rubbed York stone or Caithness flags, on a 4in to 9in (100 mm to 225 mm) hardcore or lime concrete bed with, beneath, a bedding of sand to take up the irregularity of the

1 *Discordant use of white Italian marble in an English country churchyard.*

2 *Granite setts and sandstone steps outside Winchester law courts.*

* Vandenberg, M. and Gage, M. *Hard landscape in concrete* and Handisyde, Cecil C. *Hard landscape in brick*. The Architectural Press.

stone. York stone flags have been a familiar part of most city streets since the eighteenth century, **3**. Where they survive they add much to even the simplest townscape, but the tendency to replace them with the uniformity of standard concrete slabs continues. This is particularly to be regretted in view of the free availability of York paving. In the south of England 3in (75 mm) Bath stone paving slabs, usually Corngrit or Corsham, were not uncommon. Sawn slate similarly found its way all over the country, and to a lesser extent Purbeck limestone paving. Most stone areas had suitable stone for slabbing which was often used only locally.

Road finish
A popular road finish was the granite sett, **26**. Aberdeen or Cornish granite setts of an average depth of 7in (175 mm) were laid in parallel courses, touching on a 6in (150 mm) bed of compacted hardcore and a 2in (50 mm) bed of sand, grouted in cement and sand mortar and well rammed. Current construction shown in **5**.
More kerbs have survived than paving slabs or setts. Still to be seen in as good a condition as they were a hundred years ago are the old 12in × 6in (300 mm × 150 mm) Aberdeen granite kerbs laid flat or edge bedded with top and front faces finely axed. An example of this can be seen in **3**. Decorative use of secondhand grey granite setts and new grey granite kerbs is seen in New Palace Yard, Palace of Westminster. The setts are both decorative and resistant to the wear and oil staining inevitable with the passing vehicular traffic. Sections can be lifted and replaced if necessary without difficulty for service maintenance.
Another survival testifying to durability is the beach pebble or cobble paving. Traditional paving of distinctive local character often has much charm and can still sometimes be used cheaply and effectively as pedestrian paving, **4**.
Pedestrian routes between doors and across yards are often defined by large random slabs. There were many local ways of laying this most attractive paving, but generally slabs were laid in fine gravel on a hard foundation of consolidated hardcore and rammed in. Where they were plentiful, flints were also used for paving. Nineteenth-century specifications for these were usually to pass through a 2½in (65 mm) ring.

Typical paving details
Typical paving details and steps using materials which are still available are illustrated in **5** to **17**. One important factor affecting the durability of stone paving slabs is the preparation of the base. In no circumstances should stone slabs be laid on a dense impervious foundation or jointed in dense cement mortar. Failure by spalling in freezing conditions is not uncommon where water is trapped in saturated slabs. Denser materials such as granite and flint pebbles are not susceptible in the same way, but dense jointing is still unnecessary and is liable to result in shrinkage cracking around the stones.
The general principles of base preparation should be followed as for concrete slabs, but the use of clinker is not recommended in direct contact with stone. Where clay earth is present, this may be stabilised to form a satisfactory base by removing top soil in the usual way, and spreading a layer of hydrated lime over the earth exposed. The lime is mixed with the earth for a depth of 75 mm to 100 mm in the approximate proportion of 5 per cent lime to 95 per cent earth. This base is then compacted with a roller, top dressed with 25 mm bedding sand and the slabs tapped into place. Cement-lime sand joints 6 mm wide (1:2:9) are suitable, and should be carefully filled. The practice of brushing in mortar is not recommended, especially on light coloured stone. Stains and hardened slurry are not easily removed.

3 *York stone flags in Queen Anne's Gate, London.*

4 *Random loose cobbles at Huntly Castle, Scotland.*

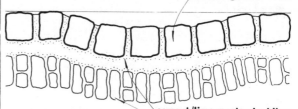
5 *Vehicular/pedestrian use: paving in granite setts showing method of forming gutter (see also **16**, **17**).*

6 *Granite setts and new grey granite kerbs in New Palace Yard, Palace of Westminster.*
7 *Setting out of radii at Palace of Westminster.*

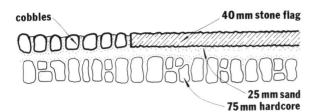

8 Pedestrian use: paving in cobbles and riven slab.

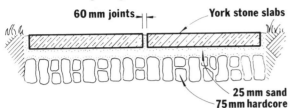

9 Pedestrian use: paving in sawn slabs (see also **3**).

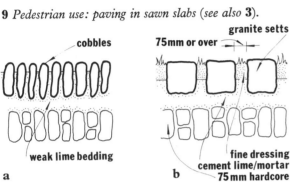

10 Pedestrian use (landscape infill). Examples of infill area paving which are uncomfortable but not prohibitive to pedestrian traffic: **a** 'horranising' or traditional Scottish form with cobbles laid 'flat parallel'; **b** granite setts, wide spaced in fine dressing (see also **11**).

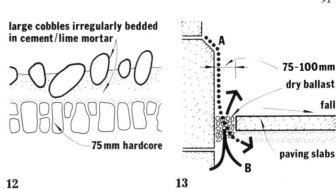

12 Anti-pedestrian use (landscape infill): graded cobbles close bedded to inhibit pedestrian use, for instance on traffic islands.
13 Pavings against buildings (see also **14**).

14 Decay in limestone retaining wall abutting granite steps (see also **17**).

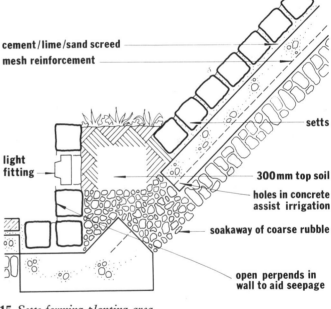

15 Setts forming planting area.

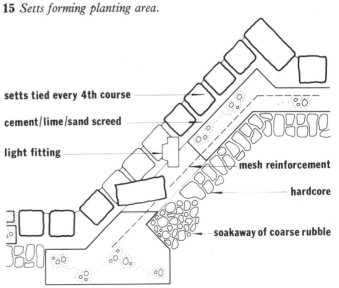

16 Sloping sett wall with integral lighting fitting.

11 Granite setts wide spaced in fine dressing, used outside car park to discourage pedestrians (see also **10**).

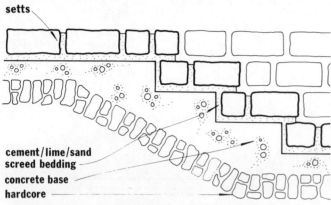

17 Setts used in paving and steps (see also **18, 19**).

18 Successful informal use of granite setts and limestone paving at the John F. Kennedy memorial at Runnymede.

19 Portland steps and granite setts at the John F. Kennedy memorial at Runnymede.

Pavings against buildings

Washings from limestone and magnesium limestone facings will deposit damaging calcium or magnesium sulphates in solution in sandstone paving slabs, causing decay.
Paving slabs abutting a stone plinth will encourage salt-bearing ground water to pass through the plinth and evaporate on the surface of the building, causing efflorescence and decay. Dense cement filling of joints will aggravate this condition. In either circumstance, it may be useful to leave a 75 mm to 100 mm margin against the wall filled with dry ballast to assist drying and drainage.
Dense granite (or concrete) steps abutting a porous limestone or sandstone retaining wall may result in the kind of decay shown in **14**. Salt-laden water from the soil is deposited in the wall and dries out. Salt crystallisation damage is accentuated by the impervious character of the paving.

Landscape walls

Masonry walls have already been discussed in this Handbook, and reference should be made to the study on retaining walls in *Hard Landscape in Concrete* which describes cavity retaining walls and concrete cantilever walls which may be faced with masonry. Correctly detailed, masonry has much to offer where large expanses of landscape wall are involved, for it will not become unsightly and will never assume the bleak quality of some uniform cast materials in unbroken areas. The risks of staining and failure through freezing should be properly considered, **20**. Seepage from soil behind a wall may cause excessive saturation with resultant frost damage. Similarly, water which has passed through soil may deposit chlorides and nitrates in the stone as it evaporates at the surface. Decay may result from this contamination, and staining will usually occur, especially where the water also seeps through concrete or lime mortar into limestone. Alkali compounds in the mortar or backing react with organic material in the limestone to form a soluble organic salt. This is seen as a dark brown discoloration on the masonry face, and may be particularly severe where rapid hardening cement has been used.

20 Frost damage to permanently saturated limestone retaining walls at Witcombe Roman villa.

The answer to this problem lies in coating the back-up wall with bituminous paint (not the back of the stone, unless there is no alternative), and in the provision of sufficient weepholes and back drainage to prevent build-up of water. Small landscape walls are now frequently built in thin slab material with rock faced edges, and these are to be seen as widely dispersed as the York stone paving, often from the same source. Frequently, however, the landscape wall is the one feature where use of local material and type of construction may come into its own, **21, 22**, and consideration should always be given to this possibility.

21 Drystone boulders of gneiss forming sheep enclosures on the Isle of Lewis.

22 *Squared chalk ashlar protected by brick footings and thatched coping, Broughton, Hampshire.*

3 Patterns in paving

● River or beach cobbles bedded in sand, stabilised earth or weak lime/sand bases, **23**, are a simple, traditional form.
● Random squared Aberdeen granite setts bedded in lime mortar give an extremely durable paving of considerable beauty, **24**.
● Of granite slabs set in and cobbled paving, the more expensive are used only on the direct pedestrian route, **25**.
● Vehicular paving in granite setts resisted solid iron-tyred and horse traffic for decades without any sign of wear, **26**.
● Squared setts are used as a roadway on the Channonry, St Machar's Cathedral, Aberdeen, **27**.
● A dramatic sweep of granite setts below the steps up to the law courts in **28**.
● Decorative infill of granite setts and trees has been used in **29**.
● The Winchester High Street pedestrian area has been paved with artificial stone. The pattern of standard squared sizes is as much a 'giveaway' of the precast nature of these slabs as is the imitation riven finish, **30**.

26 *Vehicular paving in granite setts.*

27 *St Machar's Cathedral, Aberdeen.*

28 *Winchester's new law courts.*

29 *Shopping precinct, Winchester.*

23 *Courtyard at Huntly Castle, Aberdeen.*

24, 25 *King's College Chapel, Aberdeen.*

30 *Winchester High Street: pedestrian area.*

4 Costs

In *Hard landscaping in concrete* a scale of comparative costs for materials in common use was given (see table I). The costs include bases, but are an approximate guide only. Secondhand setts or York paving may be comparable in cost with brick paving or limestone paving. The cheapest setts are currently imported from Portugal but Cornish and Irish setts are available.

Offset against the high initial cost of stone paving must be set a negligible maintenance cost, and a very high standard of appearance.

Table I Scale of comparative costs

Cost of 100 mm in-situ concrete slabs is represented as 75

	Cost
60 mm tarmac	35
50 mm precast concrete paving slab	200
cobbles set in concrete	240
brick paving	250 to 375
50 mm decorative precast 'stone'	500
York stone	550
50 mm French limestone	600

This is an approximate guide only: random rectangular slabs, for example, would be cheaper than sawn six sides, and good secondhand material could be cheaper still.

Technical study

Stone substitutes

In this short technical study, JOHN ASHURST and IAN CLAYTON discuss the use of substitutes in the repair of natural stone and give a summary of comparative costs.

1 Historical substitutes

1.01 There have been more imitations of architectural stone than any natural building material. The reasons for this persistent emulation are various, but have most consistently been for reasons of economy, availability and fashion. Today, the first of these reasons is the usual one, but substitution may be justified in a new way with the development of glass reinforced plastic and a variety of finishes where failure has occurred due to the overloading of a structure or substratum, **1, 2**.

1.02 Some of the historical substitutes for stone are now themselves the subject of conservation. The last century saw the development of a host of patent artificial stones. These were usually (in the last half of the century) compositions of Portland cement and fine aggregates, including stone dust of the type to be matched in the proportions of 2:7. Pigments were frequently included and cast blocks were sometimes carved while still green; paraffin oil and french chalk were commonly used as release agents. A few of these artificial stones were good matches, but most were too strong and subject to crazing. Rarely have they weathered in any real semblance to stone.

1.03 Most famous of the artificial stones was that developed at the Coade works on the Festival Hall site which thrived, with its many copies, for a hundred years. BRS examination of this material (which was cast in an enormous variety of architectural and landscape accessories) showed by X-ray mineralogical analysis that it was well crystallised kaolinite with some white mica. Critical to its manufacture was the control of the furnace temperature, and fluxes introduced to reduce the temperature at which vitrification occurred included crushed glass and felspar. A certain mystery still clings to the manufacturing process, but the most closely guarded secret must have been the firing, **3, 4**.

1.04 Another well known artificial stone which has enjoyed popularity from the early 16th century is scagliola, which used coloured plasters and small pieces of stone. In experienced hands, many exotic marbles were copied, and the finished, usually polished, product was frequently a spectacular achievement, **5**.

1, 2 *University College School, Hampstead. Columns restored in cast stone; frieze and dome restored in grp.*

3, 4 *Use of Coade stone in decoration around London doors.*
5 *Scagliola used in interiors of Osborne House, Isle of Wight.*

1.05 Gypsum, lime stucco, oil mastic, 'Roman cement' and coloured Portland cement renderings have all been used extensively as rendered imitations of stone, lined out in imitation of fine jointed ashlar, and sometimes heavily rusticated and cast or run in elaborate mouldings.

2 Selection of substitute

2.01 When the architect is faced with selecting a substitute for natural stone, for whatever reason, it is important to avoid a text-book approach to the problems and to opt instead for a more empirical method. The reasons for this are based on the fact that every repair job, like the structure of which it is a part, is unique; therefore rigid academic solutions can only be regarded as guidelines, not as immutable laws to be obeyed at all times.

2.02 A repair should be seen in the overall context of the structure itself. For instance, what kind of a building is it? Is it of historical importance or of no particular architectural merit? Also, the longevity of the structure should be considered. It is hardly worth while embarking on an expensive restoration programme for a structure that is going to be demolished in a few years.

2.03 When these factors have been weighed, the next consideration must be to establish exactly what causes the decay. Repair is not a cosmetic process alone; by definition it must eliminate those factors which have made it necessary in the first place if it is to be worthwhile. This may involve long, detailed consideration of the whole structure. The architect must ask himself whether the right stone was used in the first place. If so, is that stone still available? Attention should also be paid to the obvious causes of decay: the cornice not covered in lead, or a leaking gutter. This may seem elementary, but it is surprising how often the obvious is overlooked.

2.04 There are four basic methods of restoring old stonework. The most desirable for buildings of architectural merit is replacement in natural stone, **6, 7**. Stone used in the construction of very old buildings is often found to be from worked-out quarries, but this is not normally an insurmountable problem, as stone being currently cut can usually be matched up with the original. The ostensible objection to using natural stone is a purely economic one. This is not just in the area of supply and working, but because its use may involve additional shoring, lifting tackle, and strengthened scaffolds. It can also have a disruptive effect on other on-site activities. For instance, taking down a parapet before renewing a cornice can cause chaos on site and, more important, add a small fortune to the contract price.

Cast stone (sometimes known as artificial stone)

2.05 In the recent economic crisis, the most popular method of repair is probably by the use of cast stone. Precast stone is not as aesthetically pleasing as natural stone, although in some cases it has the singular advantage of superior weathering properties to many natural stones. Cast stone as an alternative to worked stone has definite economic advantages, particularly where a run is required. Once a mould is obtained of, say, a moulded cornice, one could go on turning them out ad

6 *County Hall, Lewes. Carving new stone capitals out of natural stone.*
7 *Maidenhead Bridge. Cutting out to prepare for piecing in with new stone.*
8 *York Terrace. Columns and capitals in cast stone (1971).*
9, 10 *Cast balustrade from Osborne House, Isle of Wight.*
11 *St Pancras Station. Plastic replacement of corner stone (1975).*
12 *Calthorpe Estate. Plastic repair to cornice over doorway.*
13 *Cranleigh Church, Surrey. Plastic stone mix reinforcement.*

infinitum at a highly economic rate. The difference in cost can be compared to the difference between a hand-built motor car and a mass-produced one, **8**. By the very nature of the process, the details are also more faithful than a hand-carved one as well. The moulded detail of Regency stucco was often cast in any case. However, all the arguments of shoring and contract disruption still apply for cast stone as for natural stone, and indeed the cost is very similar where there are not lots of repeats, **9**, **10**.

'Plastic' stone (artificial stone mortar used in situ)
2.06 The most economic form of stone repair, and unfortunately the one which is the most heavily criticised, is the 'plastic' stone method. Most of the criticisms of this method are misdirected. The trouble lies not with the method itself but with the often indifferent operational capabilities of those applying it, for it is a craft skill. It has the great advantage of enabling repairs to be executed with the minimum of disturbance. For example, a mullion can be repaired without disturbing the glass. Like cast stone, its weathering properties may be equal in some cases to the natural product, although its surface will never weather in quite the same aesthetically pleasing way. Of all the accepted methods of stone repair, this is the one which is closest to a cosmetic treatment, and therefore it has limitation where deep-seated structural problems are involved, **11, 12, 13**.

2.07 The mould shown in **14**, is a template cut from zinc securely fixed to a wooden mould board. Wooden rules are

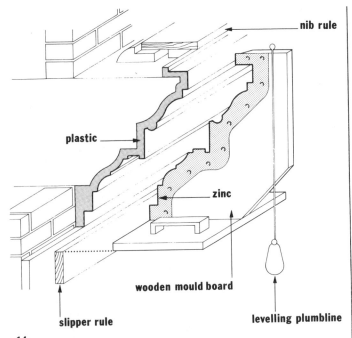

carefully fixed above and below the rough concrete core of the cornice (this might be formed in tile coursing). The core is plastered with the stucco or 'plastic stone' mix at approximately 6 mm thickness, omitting any coarse aggregate which would score and pull the wet mix off. The mould is then run along the slipper rule.

2.08 Current research at Building Research Establishment is investigating the durability and other weathering characteristics of plastic stone mixes, both traditional and with modern additives, using the natural stone as a control.

Glass fibre (grp)

2.09 A relative newcomer to the field is glass fibre. It is a particularly appropriate remedy where weight is a major consideration, and also as with cast stone where repetitive detail is required. Its use is recommended high up on a building, as it is difficult to get a convincing texture at close quarters with the present state of glass fibre technology, and there are still considerable doubts about its long-term weathering properties, **15, 16, 17, 18**. Note the replacement of cast stucco by cast grp. The constituents of the resins required to improve the flame spread qualities detract from the weathering properties. The grp referred to in table I is composed of glass fibres and an epoxy resin.

3 Weathering of substitutes

3.01 A general comment on weathering may be justified, in spite of increasing expertise in casting and reproduction in glass fibre. Natural stone, properly selected and maintained, weathers well and tends to improve in appearance with weathering. Subtle variations in tone and texture become more apparent with age. However good the cast block may be as a single unit, there is frequently no mistaking it for natural stone *en masse;* with exposure, the difference in appearance increases. Perhaps the most successful of the cast units has been as 'slates', and the least successful attempts as 'sawn' or 'riven' paving. This is partly due to more skilful handling in the former, partly to more intimate visual contact with paving.

14 *Traditional method of running a mould.*
15 *Broadcasting House, Langham Place, London. Repair using grp in* 1970.
16, 17, 18 *Kenwood House, London. Repairs of Robert Adam's stucco work in grp.*

19

19 Flamstead Church, Herts. Elevation showing clerestory windows five years after repair.
20 Detail of cast glass fibre facing to concrete backing.
21 Arundel 'Cathedral'. Unit made from glass fibre mould. Joints covered with glass fibre crockets.

3.02 'Plastic' stone, prepared by a craftsman to match decay or damage, is capable of weathering in a very similar way to natural stone, **19, 20**—provided the match has not been simply visual, but has similar properties of strength and porosity and is placed in a properly prepared cavity, anchored where necessary. However, for every good 'plastic' repair, there are a great many poor examples. At worst, a dense cement sand screed is trowelled into a ragged repair, increasing the rate of decay of the stone around it. Even if this is an increasing rarity, a large percentage of this kind of repair is still carried out with an inexpertise which becomes more apparent with years.

3.03 The finishes which can now be applied to grp are very much improved on those available ten years ago. 'Stone' finishes, where stone dust is used as the weathering surface, may be very successful, especially on high level features.

4 Costs

4.01 It is very difficult accurately to compare the cost of one method against another. Table I is only the roughest of guides. For instance, should the cornice go deeper into the wall, the price of natural stone will rise accordingly, cast stone a certain amount and the other two not at all. In any case, it might require a combination of methods. Economical lengths of run also enter into the overall cost evaluation.

Introduction to costing table

4.02 Table I sets out comparative costs of replacing natural stone with cast, plastic and grp substitutes. The case is almost impossibly hypothetical, as considerations other than cost would certainly enter into a repair analysis of this type. Further, all units in natural stone and cast stone have been replaced complete. In replacing with grp or repairing with 'plastic' stone, the exercise on the cornice has been to cut back and bolt on, or anchor and build up.

4.03 The successful use of all substitutes depends on the survival of at least one intact example, or intact run, of the original. A very badly damaged subject will be modelled in plaster and built up as necessary to provide the detail for casting.

4.04 A probable economic solution to repairing a badly damaged or decayed run of masonry of this complexity would involve the use of at least two types of repair.

4.05 The costs for natural stone vary according to type of stone, location of quarry, and workshop facilities.

20

21

Table I Comparative costs of replacing natural stone with cast, plastic and grp substitutes

Natural stone		Cast stone		Grp		'Plastic'	
Coping† Cost per metre run, for complete replacement	£40–£50	Cost is relative to quantity for complete replacement	£12	†	£15–£30	Not recommended for use in copings but still suitable for small amounts of dentistry repairs	£10
280 mm × 102 mm (11in × 4in)*							
Baluster† £20–£30 each per metre run. 3 number. Square and twisted sections will be more expensive	£60–£90	Each £18 × 3 per m. Quantity reduces cost but easiest for intricate twisted and fluted if a reasonably good model can be obtained from site	£54	†	Each £15 to £20 × 3 = £45–£60	Each £4 × 3. Not recommended if baluster has been badly decayed but suitable for small amounts of dentistry repairs	£12
153 mm × 153 mm × 610 mm (6in × 6in × 2ft)*							
Blocking course⊥ Cost is for complete replacement	£20–£25		£18		£24	Not recommended if constructional work is necessary	£12
305 mm × 153 mm (12in × 6in)*							
Cornice† Will involve work at back of parapet, cost taken per metre run	£100–£120	As for natural stone	£90	Including cutting back existing cornice	£40	Including reinforcement and rerunning	£18
686 mm × 229 mm (2ft 3in × 9in)*							
Modillion brackets† Each £20 × 3	£60	Each £10 × 3 Much cheaper than natural if carved detailing required	£30	Each £8 × 3 Sometimes an ideal use for grp owing to ease of fixing	£18	Each £3 to £6 × 3 If any quantity required it would be better and cheaper in cast or natural	£24
153 mm × 102 mm × 254 mm (6in × 4in × 10in)*							
	£361		£204		£178		£70

†Not heavy enough as a whole balustrade and would have to be strengthened with stainless steel.
*Imperial dimensions are shown as certain measurements are still taken in imperial in cases of restoration or repair to existing buildings.

APPENDIX: Trade, craft, consultative and other bodies

Aberdeen Granite Association
Investment House, 6 Union Row, Aberdeen (Aberdeen 26262).
Secretary: I. Michael S. Park. Membership among Aberdeen merchants and suppliers.

Association of Natural Stone Industries
82 New Cavendish Street, London, W1M 8AD (01-580 4041).
Secretary: E. G. Sangster. Affiliated to the National Federation of Building Trade Employers. Membership open to all companies working wholly or partly on the construction side of the stone industry. Associate membership for other companies in the stone industry. Services to members include promotion and technical facilities, consultation and advice on wages, contracts, conditions, industrial relations and training.

Building Research Establishment
Bucknalls Lane, Garston, Watford, WD2 7JR, Herts (Garston 74040; telex: 923220). Part of the Department of the Environment. Conducts study programme of stone preservation methods. Gives advice on behaviour of stone and methods of use.

College of Masons
Gen secretary: J. J. Byrne, 23 Highshore Road, London SE15.
Promotes the craft of stonemasonry and encourages the skills of apprentice masons.

Council for Places of Worship
83 London Wall, London, EC2M 5NA (01-638 0971). Formerly the Council for the Care of Churches.

Council for Small Industries in Rural Areas
35 Camp Road, Wimbledon Common, London, SW19 4UP (01-947 6761).

Department of the Environment, Directorate of Ancient Monuments and Historic Buildings
England: Fortress House, 23 Savile Row, London, W1X 2AA (01-734 6010).
Scotland: Argyle House, 3 Lady Lawson Street, Edinburgh, EH3 9SD (031-229 29191).
Wales: Government Buildings, St Agnes Road, Gabalfa, Cardiff, CF4 4YF (Cardiff 62131).
Advice on selection and maintenance of stone. (This advisory work is shared with BRE.)

Dry Stone Walling Association
Hon secretary: Miss E. M. P. Audland, Cally Estate Office, Gatehouse-of-Fleet, Scotland, DG7 2HK (Gatehouse 200). Will supply information on request about craftsmen dykers and wallers, training courses, competitions, demonstrations—but please send sae.

English Slate Quarries Association
11-12 West Smithfield, London, EC1A 9JR (01-248 6805).

Federation of Stone Industries
Alderman House, 37 Soho Square, London W1 (01-437 7107).
Secretary: D. Maxted-Jones. Amalgamating the British Stone Federation, Marble and Granite Association and Architectural Granite Association. Membership open to quarry owners and masonry firms in all sectors of the natural stone trade. Combines sectional and geographical organisation to cater for specialist and local interests within the stone industry, and receives affiliations from other groups not in direct membership. Specialist sections include stone cleaning and restoration, and quarrying.

Geological Museum
Institute of Geological Sciences, Exhibition Road, London, SW7 2DE (01-589 3444).
Geological reference library and museum. Also has a 'library' of 14 000 samples of stones, marbles, granites, etc; data available on geological characteristics.

Guild of Lettering Craftsmen
1 Park Royal Road, Acton, London W3.
Secretary: T. W. Langley.

Joint Committee for the British Monumental Industry
Alderman House, 37 Soho Square, London W1 (01-437 7107).
Secretaries: C. W. Allen, D. Maxted-Jones. Consultative body for manufacturers, wholesalers and retailers in the monumental trade. Represented on the committee are the Aberdeen Granite Association, Federation of Stone Industries, National Association of Master Masons and Scottish Master Monumental Sculptors' Association.

Letter Cutters Association
59 Rusper Road, Wood Green, London N22.
Secretary: S. Calas.

London Association of Master Stonemasons
18-20 Duchess Mews, London, W1N 3AD (01-636 3891).
Secretary: B. Luck.

Men of the Stones
The Rutlands, Tinwell, Stamford, Lincs (Stamford 3372).
Secretary: A. S. Ireson, MBE. Independent body for promoting the use of stone; offers advice on stone for old and new buildings.

National Association of Master Masons
Alderman House, 37 Soho Square, London W1 (01-437 7107).
Gen secretary: C. W. Allen. Membership of stonemasonry firms engaged in monumental and building work throughout UK.

National Fireplace Manufacturers' Association
PO Box 13, Hanley, Stoke-on-Trent, ST1 3RG Staffs (0782 29031).
Membership includes manufacturers of stone and marble fireplaces.

National Joint Council for the Monumental Industry for England and Wales
Joint secretaries:
employers: D. Maxted-Jones, Secretary, Federation of Stone Industries, Alderman House, 37 Soho Square, London W1 (01-437 7107);
operators: L. Poupard, Union of Construction, Allied Trades & Technicians, 9 Macaulay Road, Clapham, London, SW4 0QP (01-622 2362).

National Federation of Freestone Quarry Owners
c/o J. Sanderson, Heys (Britannia) Ltd, 27 Tong End, Whitworth, Rochdale, Lancs (070 685 3295).

North Wales Slate Quarries Association
Bryn, Llanllechid, Bangor, Gwynedd (Bethesda 600 656).
Secretary: H. A. Pritchard.

Northern Ireland Quarry Owners Association
c/o Jackson Andrews & Co, Chartered Accountants, River House, 48 High Street, Belfast, BT1 2BY (0232 33152).

The Orton Trust
Stoa House, Brigstock, Kettering, Northants (Brigstock 253).
Director: J. L. Nightingale. Training centre for masons, architects, sculptors and others on aspects of stone utilisation.

Royal Society of British Sculptors
8 Chesham Place, London SW1 (01-235 1467).

Sealant Manufacturers' Conference
Dickens House, 15 Tooks Court, London, EC4 1LA
For information on sources and use of joint sealants for cladding.

Scottish Freestone Quarry Masters' Association
Secretary: W. G. Paterson, Knowehead Quarry, Locharbriggs, near Dumfries (Amisfield 231).

Scottish Master Monumental Sculptors' Association
90 Mitchell Street, Glasgow C1 (041-221 0424).
Secretaries: Adam, Ker & Sangster.

Society of Church Craftsmen
c/o RWS Galleries, 26 Conduit Street, London, W1R 2TA (01-629 8300).
Secretary: Malcolm Fry.

Society of Designer Craftsmen
c/o 6 Queen Square, London WC1.

Society for the Protection of Ancient Buildings
55 Great Ormond Street, London WC1 (01-405 2646).
Secretary: Mrs M. Dance.

Standing Joint Committee on Natural Stones
Admin House, Market Square (North Side), Leighton Buzzard, Beds (05253 75952).
Secretary: D. Maxted-Jones. Consultative body on which are represented the Royal Institute of British Architects, Institution of Civil Engineers, Institute of Geological Sciences, Council for Places of Worship, Society for Protection of Ancient Buildings, Confederation on Training Architects in Conservation, Worshipful Company of Masons, Men of the Stones, The Orton Trust, Federation of Stone Industries, National Association of Master Masons, Union of Construction Allied Trades & Technicians, the DOE Directorate of Ancient Monuments and Historic Buildings, with observers from Department of the Environment.

Standing Joint Committee on Recruitment and Training of Architects for Care of Old Buildings
44 Queen Anne's Gate, London SW1.

Worshipful Company of Masons
9 New Square, Lincolns Inn, London WC2 (01-405 6333).
Clerk: H. J. Maddocks, City of London Livery Company. (Master for 1974-75 is H. L. V. Lobb, CBE.)

Yorkshire Quarry Owners Federation
10 Bradford Road, Brighouse, Yorks (Brighouse 3457).
Secretary: B. Sutcliffe.

Acknowledgement

This list was selected from Stone Industries' *Natural stone directory* 1974 (reference 1).

Acknowledgements

Acknowledgements are due as follows for illustrations:
Geology: Technical Study 1: 2–6, 8 Institute of Geological Sciences 7 Francis G. Dimes 9 R. H. Roberts
Geology: Technical Study 2: 1, 4 *The Architects' Journal* (Bill Toomey) 2, 5a Institute of Geological Sciences 3 Henk Snoek 5b R. H. Roberts 6a, 6b Dr F H Broadhurst 7 *The Architectural Review*
Geology: Technical Study 3: 1 Francis G. Dimes 2 Eric de Mare 3 Dr F. H. Broadhurst 4 *The Architects' Journal* (Sam Lambert) 6–8, 11 Institute of Geological Sciences 10 *The Architectural Review*
Geology: Technical Study 4: 1 Francis G. Dimes 2–4, 7 Institute of Geological Sciences 5 *The Architects' Journal* (Norman Gold) 6 *The Architectural Review* (R. Einzig)
Quarrying: Technical Study 1: 1 John Ashurst 2–4 Stone Industries
Quarrying: Technical Study 2: 1–5, 7–9 *The Architects' Journal* (Bill Toomey) 6 Stone Industries
Processing and Site Works: Technical Study 1: 1, 8, 13–15 *The Architects' Journal* (Bill Toomey) 2–7, 9–12 Stone Industries
Processing and Site Works: Technical Study 2: 1 Natural Stone Quarries (James Riddell) 3, 4 Stone Firms Ltd, Corsham, Wiltshire 5 *The Architects' Journal* (Bill Toomey)
Stone Training: Technical Study 1: 1–9 *The Architects' Journal* (Bill Toomey)
Maintenance: Technical Study 1: 1–11, 15, 20, 21 Reg Wood 12 F. G. Dimes 16a S. Parker, Stoneguard, Ruislip, Middlesex 17–19 Corinne Wilson 22 John Ashurst 23 Building Research Establishment
Maintenance: Technical Study 2: 1–15 Reg Wood
Maintenance: Technical Study 3: 2–5, 8–12 Building Research Establishment
Specification: Technical Study 1: 1–4 Building Research Establishment
Specification: Information Sheet: 1: 2–4 Building Research Establishment
Stone in Use: Detail Sheet 10: 1, 3, 4 Reg Wood 2 Gillian Carter
Detail Sheet 18: 1 Chichester Cathedral Works Organisation (Thomas Photos)
Hard Landscape in Stone: 24–27 Department of the Environment
Stone Substitutes: 8 Natural Stone Quarries 19, 20 Reg Wood; other photographs supplied by John Ashurst, Szerelmeys Ltd, and the Crafts and Lettering Centre.

All Institute of Geological Sciences photographs are reproduced by permission of The Director, and are NERC copyright.

Index to stones

This index is to help architects select from available stones. It is not possible for the handbook to illustrate the subtle differences of colour and pattern, the variations in local use, and the weathering characteristics for hundreds of stones. But to go some way towards this, section 1 'Geology' lists many stones with their characteristics and, when possible, where examples of their use can be found. However information in section 1 is classified geologically; here the index lists the stones alphabetically so that architects can discover where the appearance and durability of a particular stone may be seen at first hand. (Further information can be gained by visiting the quarries. For their location see the quarries directory—listed in the general index.)

A

Aberdeen granite 7
Alabaster 17
Alps grey granite 19
Alta quartzite (altazite) 19
Amberley stone 16
Ancaster stone 16

Anston stone 15
Arabescato 19
Arkose 3
Auchinlea stone 13
Aurisina 17
Australian 19

B

Babbacombe 15
Ballachulish slates 19
Balmoral granite 7
Bardiglio 19
Barge quartzite 19
Barnack stone 16

Basalt 8
Base bed 16
Bath stone 16, 88

Baveno granite 7–8
Beer stone 16
Bethersden marble 16

Birchover gritstone 12

Black granite 9
Blaxter stone 12

Bleu belge 17
Blue lias 16
Blue pearl 6
Bolsover Moor 15
Bolton Wood stone 13

Bonaccord black 9
Bonaccord red 7
Box Ground stone 16

Bradford stone 16

Bramley Fall 12

Brathay slate 19
Breche rose 19
British marbles 19
Brocatello 17
Broughton Moor 19
Burlington slate 19
Burwell stone 16
Buttermere 19

C

Caen stone 17
Caithness flagstone 12
Campan 19
Carboniferous Limestone 12, 14
Carrara marble 17, 19

Casterton stone 16

Chalk 16, 93
Channel Isles 7
Charlwood stone 16

Chilmark stone 16
Cipollino 19

Clashach stone 13
Clipsham stone 16, 87

Clunch 16

Coal Measures 12–13
Colleyweston slate 16
Combe Down stone 16

Connemara marble 19
Corrennie granite 7
Corsehill stone 13
Cotswold slates 16

Craigleith stone 12

Crossland Hill stone 12

D

Dale stones 12

Darley Dale stone 12

Deepdale 15
Delabole slate 19
Derbydene 15
Derbyshire Fossil 15
Devon and Cornwall granite 6
Devon marble 15
Diamantzite 19
Diamond Black granite 9
Diorite 1, 8
Doddington stone 12
Dolerite 8
Doulting stone 16

Dukes Red 15
Dunhouse stone 12

Duston stone 13

E

Ebony Black granite 9
Elland Edge flagrock 13

Elland Edge stone 12

Elterwater 19
Elvan 8
Ematita 'granite' 19
Emerald pearl 8–9

F

Flint 17, 89, 90
Forest of Dean stone 13

Frosterly marble 15

G

Gabbro 8
Gneiss 18, 19, 92
Granite 1, 6–7, 23, 26–7, 89, 90, 91, 92
Granodiorite 1, 8
Great Oolite 16
Green Brae 13
Greensand (Lower) 16
Green Ventnor 13
Grey Royal 7

Grinshill stone 13
Gritstone 3, 12
Guiting stone 16

H

Hadene stone 15
Hailes stone 12

Hall Dale stone 12

Ham Hill stone 16
Hereford stone 12

Hollington stone 13
Hopton Wood stone 15
Hornton stone 16
Horsham stone 11, 12
Huddlestone stone 16

I

Imperial Mahogany granite 8
Imperial Pearl 9
Imported marbles 19
Inferior Oolite 16
Iona marble 19
Ipplepen 15
Israel marble 17
Italian quartzite 19

J

Jaune lamartine 19

K

Kemnay granite 7
Kentish Ragstone 16

Ketton stone 16

Kirkstone Green 19

L

Lakeland Green 19
Lepine 17
Large Paludina marble 16

Larvikite 8
Laughton stone 16

Lazonby Red 13
Lias 16
Light Pearl 8–9
Limestone 3, 13–17, 23, 91, 92, 93
Lincolnshire Limestone 16
Locharbriggs stone 13
Longchant stone 17
Lower Greensand 16
Luxullianite 6

M

Madreperla 19
Magnesian Limestone 15
Mansfield stone 13
Mansfield Woodhouse stone 15
Marble 3, 17, 18–20, 23, 89
Millstone grit 12
Monks Park stone 16

Moulin à vent 17
Moulton 13

N

Nabresina 17
New Red Sandstone 13
Norwegian Rose 19

O

Ogwell 15
Old Red Sandstone 12
Onyx marble 17
Oolite 3, 16
Orton Scar 15
Otta slate 19

P

Paludina marble 16
Parian marble 19
Pavonazzo 19
Penmon limestone (marble) 15
Pennant stone 13

Penrith Red 13
Pentelic marble 19
Pentewan stone 8
Peterhead granite 7
Petit Granit 6, 17
Petit Tor 15
Petworth marble 16

Pillauguri slate 19
Polyphant stone 8
Porphyry 1, 8
Portland stone 5, 15, 16, 83, 84, 92
Prudham 12
Pudding stone 13
Purbeck marble 16
Purbeck Portland 16

Q

Quarr stone 16
Quartzite 18, 19, 23

R

Rainhill stone 13
Red Wilderness stone 12
Rembrandt 19
Richemont 17
Roach 16
Roche Abbey stone 15
Roman stone 17
Rose Brae 13
Rose Swede granite 7
Ross of Mull granite 7
Rouge Griotte 17
Rouge Royal 17

S

St Aldhelm 16

St Anne 17
St Bees 13
St Boniface 13
St Maximin 17
Safari quartzite 19
Salterwath 15
Sandstone 1, 3, 10–13, 89
Sanfont stone 19
Sarsen 12, 13
Schist 3, 18, 19, 20

Scottish granite   7
Serpentinite   1, 8, 18, 19, 20
Shap granite   6–7
Siena marble   17
Silver White granite   8

Skye marble   19
Skyros   19
Slate   3, 4, 18, 19, 20, 23, 80, 86
Small Paludina marble   16

Spinkwell stone   12

Spoutcrag   19
Spynie stone   13
Stainton stone   12

Stancliffe stone   12

Steetly stone   15

Stoke Ground   16

Stonesfield slates   16

Stowey stone   16
Sussex marble   16

Sussex sandstone   13
Swaledale Fossil   15
Swedish Green   19
Swedish Green granite   8
Syenite   1, 6, 8

T

Tadcaster   16
Taynton   16

Tinos   19

Tisbury stone   16
Totternhoe stone   16
Travertine   3, 17, 86

V

Verde antico   19
Verde ematita   19
Verde fraye   17, 19

Virgo granite   7
Verona   17

W

Weald Clay   16
Wealden Sussex stone   13

Weldon stone   16

Wellfield stone   12

Welsh slates   19
Westmorland Green   19

Westwood Ground   16

Whatstandwell   12

Whit Bed   16
Winsley Ground   16
Woodkirk stone   13

Woolton   13

Y

York stone   12, 90, 93

General index

A
Acrylics 61
Algae, causing decay 43
Alkoxysilanes 61
Ashlar
  —definition 29, 71
  —bonded walls 71, 85
  —to masonry walls 74
Artificial stones 56–7, 94–9
Attrition 43

B
Basalt 8
Bedding 10, 25, 26, 55, 66
Biological inhibitors 50
Blasting 26
  —abrasive 46
  —methods 46–7
  —wet 47
Block selection 27
Bonaccord black granite 23
Building crafts training scheme 36, 39

C
Carboniferous limestone 12, 14
Carboniferous sandstone 10, 12–13
Cement mixes 54, 66, 90
Checklist for repair 67
Chemical tests 62
Clastic 10, 13
Cladding 33–35, 71, 74–80, 83, 85, 86, 88
  —compression beds in 35, 71, 76
Cleaning 40–57
  —abrasive blasting 46–7
  —case for 44
  —chemical 47, 49
  —high pressure lance 46
  —mechanical 48
  —methods of 44–5, 49
  —problems 48
  —recent developments 49, 58–61
  —steam 46
  —washing 45
  —water penetration 45
  —wet blasting 47
Clipsham stone, durability 63
Coade stone 94
Cobbles 89–93
Column bases 62
Conductivity 69
Consolidants 50
Consultative and other bodies 100
Copings 79
Craft bodies 36, 38, 39, 100
Creeper, removal 53
Crystallisation test 63–4
Cutting, cutting out 54, 55

D
Damp proof courses 66
Decay agents of 40
  —chemical 41
  —organic 43–4
  —physical 43
  —testing 82–4
Decorative uses 87, 88
Diamond Black granite 9
Dirt, removal 53
Dimension stone 1
Dolerite 8
Drawings of stonework 81, 82
Durability factor 63

E
Epoxy resins 35, 57, 61, 86
Exposure 69
Extraction 24, 26

F
Feldspar 1, 6, 7, 8
Finishes 9, 31, 32, 66
Fixing 33–5, 66
  —ashlar under 75 mm 78, 85, 86, 88
  —ashlar over 75 mm 77, 83, 85
  —costs 34
  —loadbearing 34, 77, 78, 83, 86, 88
  —location 76
  —methods 33
  —restraint 34, 77, 78, 83, 86
  —tolerances 34
Flagstone 10, 12–13, 89, 90, 93
Freezing tests 63
Frost damage 43, 50, 90, 92
Fungi, causing decay 43

G
Gabbro 8
Galetting 54
Geological column 5
Geological museum 21
Glass fibre (grp) 97–9
Gneiss 6, 18, 19, 20
Granite 1, 3, 6–7, 23, 89, 90, 91, 92
  —Aberdeen 7
  —availability 7, 22, 26–7
  —Channel Isles 7
  —Corennie 7
  —definition 6
  —Devon and Cornwall 6
  —extraction 26
  —imported 7–8
  —Kemnay 7
  —Peterhead 7
  —quarrying 26
  —Ross of Mull 7
  —Scottish 7
  —Shap 6–7
  —weathering 1
Granodiorite 8

H
Handworking 32, 37, 38, 39, 57
Hard landscaping 89–93
Heat loss 70
Heat transfer 69
Horsham stone 11, 12

I
Igneous rocks 1, 6–9
  —classification 1
  —imported 7–9
  —joints 4
Impregnation
  —in-depth 60–1
  —results of 61

J
Joints 4, 24, 26, 66
  —joggles 56, 78
  —repointing 54, 67
  —secret key joint 72

L
Landscaping 89–93
Lathes 30
Lichens, causing decay 44
Limestone 3, 13–17, 23, 91, 92, 93
  —chemical 13
  —durability 63
  —imported 17
  —mineral content 13
  —names, sources, characteristics 15–17
  —organic 13
  —quarrying 24
  —slates 25
  —types 14, 15–17
Loadbearing members, strength 62
Luxulianite 6

M
Machining 30–31
Marbles 3, 17, 18–20, 23, 89
  —imported 19, 20–1
  —names, sources, earacteristics 19
Metamorphic rocks 18–21
  —classification 3, 19
  —joints 4
  —names, sources, characteristics 19
Mica 1, 18
Microscopy 3, 63
Millstone grit 12
Moisture
  —content 40, 69
  —treatment 50, 53
Mortar mixes 54, 66
Mosses, causing decay 44
Mould cutting 27, 96–7

N, O
New Red Sandstone 13
Old Red Sandstone 12
Onyx definition 17
Onyx-marble, definition 17
Organic growth 43

P
Paving 12, 13, 15, 16, 19, 29, 89–93
Pentewan stone 8
Planers 30
Plastic stone 56, 57, 96, 99
Plinth courses 62
Polishing machines 30, 31
Pollutants 40, 41, 42
  —atmospheric 41
  —soluble salts 40, 41, 47, 50, 58–61
  —water 40
Polyphant 8
Pores, size and distribution 62–3
Porosity 62, 69
  —measurement 64
  —testing 62, 64
Porphyry 8
Poulticing 58–61
Pre-tensioned panels 35
Primary sawing 28

Q
Quarries 22, 100
  —British 22
  —distribution 2
  —future 23
  —imports 23
  —ownership 22

Quarries directory 23
Quarrying 24
  —methods of 24–7
Quartz 1
Quartzite 18, 19, 23

R
Repair 51–7, 94–9
  —anchorages 56
  —checklist 67
  —colouring 55
  —defects 53
  —epoxy resin 35, 57, 61
  —piecing in 54, 55, 56
  —plastic stone 56
  —policy 53
  —re-dressing 56
  —veneering 56
Repointing 54, 67
Retaining walls 91–2
Roads 90
Rocks 1–21
  —age 5, 12–13, 15–16
  —classification 1, 3–4
  —distribution 2, 22, 100
  —formation 1, 3–5
  —groups 1, 3, 12–13, 15–16
  —igneous 1, 3, 6–9
  —metamorphic 3–4, 5
  —naming 6
  —sedimentary 1, 3, 10–17
  —joint systems 4, 24, 66
Routing machines 30

S
Salts, damage 40, 41–3, 50, 52, 58–61, 63–4, 66, 92
Sandstone 1, 3, 10–13, 89
  —imported 12
  —mineral content 10
  —names, sources, characteristics 12–13
  —quarrying 24–6
  —types 10–13
Saturation coefficient 64
Saw/grinder 30
Saws 28–30
  —types 28
  —uses 28
Schist 3, 18, 19, 20
Secondary sawing 29
Secondhand stone 54, 55
Serpentinite 18, 19, 20
Sedimentary rocks 1, 3, 4, 10–17
  —classification 3
  —distinctive stones 17
  —joints 4, 24
Setts 89–93
Silicones 49
Site works 33
Slate 3, 4, 18, 19, 20, 23, 80, 86
  —dressing 31
  —extraction 26
  —limestone 24–5
  —roofing 80
  —splitting 26–7, 29
Sodium sulphate crystallisation test 63–4
Solvents in impregnation 61
Soffits 79
Specifications 27, 65–7
Specification clauses 65–7
Splitting machines 31
Staining 33, 48, 66
Steps 89–93
Storage and handling 32, 66
Stone
  —availability 65, 100
  —defects 54–5

—drying out 25–6
—pre-assembly 35
—protection 33
—seasoning 25–6
—selection 65, 89
—splitting 25, 26–7, 29, 31
—storage and handling 25–6, 32, 66
—strength 62, 65
—substitutes 56–7, 94–9
—testing 62–4, 65
—use
  —buildings 1, 81–8
  —decorative 87–8
  —details 71–80
  —landscaping 89–93
—veneers 35
Stone buildings maintenance 51–7

Stonework drawings 81, 82
Strength testing 62, 65
Structural defects 51, 53
Surface treatments 40–50
  —recommendations 50
Syenite 8

T

Templates 39, 96–7
Tertiary sandstone 12, 13
Testing 62–4, 65
Thermal resistance 69
Thermal stresses 43
Thermal transmittance 68–70
Tolerances 34, 66

Trade bodies 100
Training schemes 36–9

U, V

U-values 68–70

Vegetation, removal 53
Veneers 35, 56

W, Y

Walls
  —bonded ashlar 71, 85
  —concrete, stone faced 76

—masonry, ashlared 74, 75
—rubble core 72
—solid rubble 73
—solid stone 73, 81, 82, 84, 85
—thermal transmittance 68–70
Washing 44, 45–6, 47, 48
Waste, use 24
Water absorption 63, 64
Waterproofing 49–50
Water repellants 49–50
Weathering 40–4, 97
  —sedimentary rocks 1
Wind damage 43
Windows, solid stone 72
Workshop practice 28–32

York stone 12–13, 90–3